SEEKING JUSTICE I
SACRIFICE ZONE

Seeking Justice in an Energy Sacrifice Zone is an ethnography of the lived experience of rapid environmental change in coastal Louisiana, USA. Writing from a political ecology perspective, Maldonado explores the effects of changes to localized climate and ecology on the Isle de Jean Charles, Grand Caillou/Dulac, and Pointe-au-Chien Indian Tribes. Focusing in particular on wide-ranging displacement effects, she argues that changes to climate and ecology should not be viewed in isolation as only physical processes but as part of wider socio-political and historical contexts. The book is valuable reading for students and scholars in the fields of anthropology, sociology, geography, environmental studies and disaster studies as well as public policy and planning.

Julie K. Maldonado is Director of Research for the Livelihoods Knowledge Exchange Network (LiKEN) and a lecturer in the Environmental Studies Program at the University of California-Santa Barbara, USA. She works with the Institute for Tribal Environmental Professionals and co-facilitates 'Rising Voices: Collaborative Science with Indigenous Knowledge for Climate Solutions'. Her work focuses on climate adaptation, disasters, displacement, resettlement, and environmental and climate justice.

SEEKING JUSTICE IN AN ENERGY SACRIFICE ZONE

Standing on Vanishing Land in Coastal Louisiana

Julie K. Maldonado

Routledge
Taylor & Francis Group

NEW YORK AND LONDON

First published 2019
by Routledge
52 Vanderbilt Avenue, New York, NY 10017

and by Routledge
2 Park Square, Milton Park, Abingdon, Oxon, OX14 4RN

Routledge is an imprint of the Taylor & Francis Group, an informa business

© 2019 Taylor & Francis

The right of Julie K. Maldonado to be identified as author of this work
has been asserted by her in accordance with sections 77 and 78 of the
Copyright, Designs and Patents Act 1988.

Library of Congress Cataloging-in-Publication Data
A catalog record for this book has been requested

ISBN: 978-1-62958-400-3 (hbk)
ISBN: 978-1-62958-401-0 (pbk)
ISBN: 978-1-351-00294-3 (ebk)

Typeset in Bembo
by Apex CoVantage, LLC

To the members of the Isle de Jean Charles, Grand/Caillou Dulac, and Pointe-au-Chien Indian Tribes, for all the days and years you have spent sharing stories, meals, laughter, and sense of community with me. Your fortitude, good humor, and celebratory nature in the face of extreme adversities are a gift and inspiration to all those whose hearts and lives you touch.

All royalties from this book will be donated to the Isle de Jean Charles, Grand Caillou/Dulac, and Pointe-au-Chien Indian Tribes.

CONTENTS

FIGURES

ACRONYMS

API	American Petroleum Institute
BCCM	Biloxi-Chitimacha Confederation of Muskogees
BP	British Petroleum
BTNEP	Barataria Terrebonne National Estuary Program
BIA	Bureau of Indian Affairs
CEO	Chief Executive Officer
CO_2	Carbon dioxide
CPRA	Coastal Protection and Restoration Authority
DFDR	Development-caused forced displacement and resettlement
EIS	Environmental Impact Statement
EPA	Environmental Protection Agency
FEMA	Federal Emergency Management Agency
GO FISH	Gulf Organized Fisheries in Solidarity and Hope Coalition
IOM	International Organization for Migration
IPCC	Intergovernmental Panel on Climate Change
LMOGA	Louisiana's Mid-Continent Oil and Gas Association
NASA	National Aeronautics and Space Administration
NOAA	National Oceanic and Atmospheric Administration
RESTORE Act	Resources and Ecosystems Sustainability, Tourist Opportunities and Revived Economies of the Gulf Coast States Act
SLFPA-E	Southeast Louisiana Flood Protection Authority-East
UNFCCC	United Nations Framework Convention on Climate Change
UNDRIP	United Nations Declaration on the Rights of Indigenous Peoples
US	United States
USACE	United States Army Corps of Engineers
USDA-NRCS	United States Department of Agriculture's Natural Resources Conservation Service
USGCRP	United States Global Change Research Program

ACKNOWLEDGMENTS

This book is just one part of a journey that I have been honored to share with so many people. To my bayou family, words cannot do justice to what our time together has meant to me. It has been a true privilege and gift to share in your stories, laughter, and tears and to feel like part of your greater family. My life has been forever enriched by your friendship and of course the tastes of all the food you so generously share. A special thank you to Chief Albert Naquin for his dedication, gentle guidance, perseverance, and sense of humor; to Theresa and Co-Chairman Donald Dardar for their passion, energy, and making sure I always had plenty to eat; to Chief Shirell Parfait-Dardar for her determination, strong voice, and hugs; to Chris Brunet for his patience, laughter, and long afternoons spent together enjoying the breeze; to Marlene Foret for her resolve and kind spirit; to Chairman Chuckie Verdin for his leadership and fortitude; to Crystlyn Rodrigue for our late night talks and sharing in life's little pleasures; to Babs Bagwell for our life-story sharing and nighttime walks along the bayou; to Wenceslaus and Denicia Billiot and Maryline Naquin, for sharing your knowledge, wisdom, and kindness over many coffee hours; to Jake Billiot for showing me the ins and outs of shrimping life; to Chantel Comardelle for her dedication and passion; to JR Naquin, for his friendship and with whom I would happily drive many more thousands of miles; to Cheril Dupre for always making sure I had a roof over my head; and to Rosina Philippe, who first showed me the beauty of the bayou and for welcoming me into your family.

A very special thank you to Kristina Peterson, without whom I would not have experienced the joy of becoming part of the bayou family; your guidance, wisdom, and passion know no bounds. A special note of gratitude to Shirley Laska for sharing her years of experience and advice; to Richard Krajeski, who taught me to question everything and always find new ways to ask a question; to Rebecca and Jason Ferris, for their vision and responsible, collaborative work; to Nathan Jessee for his energy and thoughtfulness; and to Alessandra Jerolleman, who is a true model for what it means to fully dedicate one's self to service for a just cause. I cherish your advice, encouragement, honesty, and friendship.

I am profoundly grateful to Brett Williams, Dolores Koenig, and David Vine for their critical feedback, encouragement, and support over the years, including in seeing me through my dissertation work, from which this book emerged; to Michael Cernea for sharing his unmatched wealth of knowledge and expertise to help guide my graduate studies and beyond; to Dvera Saxton, Abigail Conrad, Jennifer Delfino, Matt Thomann, and Nell Haynes for our virtual hangouts to see each other through our research and writing; and to Sara Moore, without whom this work would still be only a tangled mess on a computer screen. Thank you to American University, the Public Entity Risk Institute, and the Natural Hazards Center at the University of Colorado, Boulder for funding and supporting the research that informed this book.

I am deeply thankful to my friends, mentors, colleagues, co-writers, and co-imaginative forces whose words, work, and wisdom provide endless inspiration and have influenced this work: Ryan Alaniz, Roberto Barrios, Paulette Blanchard, Robin Bronen, Katherine Browne, Karletta Chief, Michelle Companion, Susan Crate, AJ Faas, Shirley Fiske, Ava Hamilton, Susanna Hoffman, Katharine Jacobs, Kirk Jalbert, Barbara Rose Johnston, Kathy Lynn, Elizabeth Marino, Shannon McNeeley, Beth Rose Middleton, Anthony Oliver-Smith, Laura Olson, Stephanie Paladino, Simona Perry, Denise Pollock, Dana Powell, Margaret Hiza Redsteer, Mark Schuller, Jeanne Simonelli, Kalani Souza, Jean Tanimoto, Bill Thomas, Melissa Watkinson, Kyle Whyte, Dan Wildcat, and many others who have gifted me with their friendship and teachings over the years.

I remain always thankful to the Rising Voices 'ohana/family, whose teachings are too numerous to relay here. You inspire me to be a better scholar, advocate, ally, and overall human being. In particular, to Bob Gough, whose gentle guidance has been a force of nature and wisdom for which I am forever grateful. A special note of deep appreciation for Heather Lazrus, whose partnership, friendship, and holding my hand through this journey are among my most cherished gifts; your spirit, vision, and leadership are a continuous source of inspiration. And to my leftover believers, whose friendship and unwavering belief give me the strength and courage to keep on.

I am especially grateful to my Livelihoods Knowledge Exchange Network co-conspirators, Betsy Taylor and Mary Hufford, who make imagining and enacting change in this world both possible and fun; to my colleagues at the Institute for Tribal Environmental Professionals (Ann Marie Chischilly, Nikki Cooley, Karen Cozzetto, and Susan Rose) for your resolve and initiatives to support communities to shape their futures that are grounded in and guided by tribal sovereignty, wisdom, and cultural practices and beliefs; to my scholar-activist colleagues at the University of California, Santa Barbara, you emulate the transformative role that radical scholarship can play in climate and environmental justice; and to my students, who are a constant reminder that this world is being placed in caring and capable hands. I am truly indebted to the leaders of the Ogoni Solidarity Forum, particularly Pastor

Barry and Dora Wuganaale, who taught me what it takes to continually stand for justice; your resolve and hard work go unmatched.

A very special thank you and an ocean of gratitude to my parents, Sylvia and Rick, for their unconditional love and confidence, continued support, and unwavering belief in me. Most importantly, to my husband Phil, who makes me feel both lucky and good; to say patience is a virtue barely touches the surface. There is no one I would rather take this journey with than you. And equally as important, to my daughters Jaden and Sierra, you are my light. When you wrap your arms around me, I know it is all worth it.

INTRODUCTION

"It was like paradise." We stood at the end of the road. He wore his Native Veterans hat and that look on his face where he is about to tell a joke, a jest of seriousness that immediately turns into a beaming smile. Chief Albert, Traditional Chief of the Isle de Jean Charles Tribe, took a deep breath. The laughter stopped. "My grandpa started it and it looks like I'm going to finish it."[1] He pointed toward the south where he grew up, where he spent his childhood days drying muskrat skins and shrimp. He described the land and woods that once existed for miles south of his family's house. Now, the house was gone, taken by a storm. The little sliver of land that remained behind where the house once stood quickly turned into eroding marsh. The rest vanished under the salty waves.

★ ★ ★

Driving south down the bayous of coastal Louisiana for the first time, I saw a sweeping landscape of interwoven water and land, dotted with the gray, bare skeletal remains of trees; the limbs pointed to areas that once were land but now were water. I passed scattered houses and trailers: some empty, some occupied, some elevated 19 feet to avoid flooding, and some on the ground and in danger. The first time I drove down the Island Road, a narrow strip of asphalt filled with cracks that connects Isle de Jean Charles to Pointe-au-Chien, I could see the summer heat radiating off the road. There was no one else in sight. I stopped in the middle of the two-mile stretch and sat in the car, staring in the rearview mirror at the dilapidated road behind me, the stretch of road ahead, and the vast water surrounding me mere feet from either side of the car. I saw what others who had been south/down the bayou for the first time described to me as feeling like "the end of the world," or what Roy Nash, Special Commissioner of the Bureau of Indian Affairs, once called "the strangest colony on the American continent."[2]

At the time, I could not imagine the web of family, tribal, and colonial relationships that intertwined with the water and the vanishing land to shape coastal lives. At first this place seemed motionless. It was still, silent, and yet, if I sat and listened, I could almost hear the change. If wisdom does indeed sit in places,[3] I had to rest my eyes upon this place, upon the dead trees dotting the landscape as ghostly reminders of what the land once looked like, sit with its people underneath a raised house enjoying the afternoon breeze, and just listen.

When I moved to the region two years later and residents allowed me into their homes and communities, this intricate web—punctuated by individual inhabitants and stories of life on the bayous—became visible. Among the most striking stories was that of the relationship between the government, the oil and gas industry, the experiences of increasing flooding, hurricanes, and encroaching waters, and the consequential decimation of coastal land, culture, and ways of life.

Louisiana's Bayou Country

Home to over two million people and containing the seventh-largest delta in the world,[4] coastal Louisiana is an elaborate network of diverse habitats and landforms, including natural levees, ridges, barrier islands, forested wetlands, and marshes connected together in an ecosystem of deltaic plains formed over thousands of years from deposits of sediment, soil, and clay from the Mississippi River.[5] As Kerry St. Pé, the Executive Director of the Barataria-Terrebonne National Estuary Program (BTNEP) explained,

> Different ecologically than most of the world because we live at the end of one of the world's great rivers. And the morphology that's due to the annual flooding of the Mississippi River and when we seek high lands to build our homes and communities we go to the water. . . . We depend on the webbing that's the wetlands to protect these communities, that's what's protected us for generations, hundreds of years.[6]

However, the intricate web of diversity that makes up the local ecosystem is unraveling. Louisiana's coastal communities' cultures and water-based settlements and livelihoods are threatened by rapid and extreme land loss and coastal erosion due to co-occurring adverse events—oil and gas extraction, changing waterways and other development projects, oil disasters, increased exposure to hurricanes and flooding, sediment subsidence, and sea level rise. The evidence is seen most explicitly within the settlements of coastal tribes in Louisiana.

Hiding from government-ordered relocations in the late 1700s and first half of the 1800s, Native people established settlements in the previously uninhabited southern ends of coastal Louisiana's bayous—referring to a slow-moving stream that flows back and forth as the tide goes in and out.[7] Coastal tribes had a wealth of ecological and economic resources, including barrier islands, extensive life-giving estuaries, and an abundance of fishing resources.

Among the tribal communities in coastal Louisiana, the Isle de Jean Charles and Grand Caillou/Dulac Bands of Biloxi-Chitimacha-Choctaw Indians and Pointe-au-Chien Indian Tribe—three small, state-recognized tribes with settlements located on narrow strips of land in coastal Louisiana,[8] approximately 80 miles southwest of New Orleans—are historically fishers, trappers, farmers, and hunters and include descendants of Biloxi, Chitimacha, Choctaw, Acolapissa, and Atakapa Indians.[9] The three Tribes have common ancestral connections, yet each one traces its heritage back to a specific set of Native, Acadian, and French ancestors that established each individual settlement. They are located in Terrebonne and Lafourche Parishes (similar to a county),[10] the heart of Louisiana's "Cajun country."[11] The Parishes are among south Louisiana's seafood, agriculture, and oil hubs.

The Tribes are characterized by their food practices, oral history, mutual aid, and trading resources, such as trading shrimp for crab or oysters or helping each other with boat maintenance or housing construction. Many families still maintain partial subsistence livelihoods, but much has been lost. Changing climate and landscape patterns have left the coastal tribes facing significant challenges and losing cultural, medicinal, subsistence, and environmental resources. Yet, despite the ever-increasing layers of disasters, the element that most bubbles to the surface is the spirit of the people to maintain their sense of humor and generosity and to continue to adapt, resist, and persevere.

The Lived Experience of Rapid Environmental Change

Hydrological, meteorological, and environmental disasters, extractive industries, river mismanagement, and anthropogenic climate change[12] are drastically transforming coastal Louisiana's near-shore and landscape. Loopholes in environmental regulations and the privatization of lands and waters have enabled environmental destruction and put communities in harm's way. The colonial-initiated capitalist drive for power and economic gain that envisions Earth's lands, waters, and minerals as commodities to be controlled, owned, exploited, and sold to the highest bidder has transformed places like coastal Louisiana into what political scientist Herbert Reid and anthropologist Betsy Taylor (2010: 11) called "new geographies of domination."

Based on ethnographic research methods, this book investigates the lived experience of rapid environmental change engineered through a specific socio-historical and political foundation. To do so, I examine the effects of the climate change–disaster–energy nexus on the Isle de Jean Charles, Grand Caillou/Dulac, and Pointe-au-Chien Indian Tribes. The research includes a particular focus on the resulting displacement effects, with an understanding that "displacement" constitutes more than physical human movement; people with deep ties to place can experience certain effects of displacement while still physically in place.[13]

The book's analysis is grounded in a political ecology perspective, which includes what anthropologist Brett Williams (2001: 409) defined as "the power relations, inequalities, connections, and contradictions that join natural and social processes over time." The book aims to demonstrate that climate change and ecological shifts cannot be understood only as physical processes, but must be considered within specific socio-historical and political contexts that have marginalized select communities.

This book is not about one event or disaster, but the accumulation of co-occurring adverse events and habitual disasters and injustices over time. Such events highlight local pre-existing socio-economic inequalities and reproduce those inequalities as the disaster unfolds.[14] As opposed to focusing on the physical properties, I start from the understanding that disasters represent "complex combinations of natural hazard agents and human action."[15] As such, there is no such thing as a *natural disaster*. In sum, the book illustrates how the contemporary capitalist-driven political-economic structure that emerged from the colonial settler process has manufactured the vulnerability of coastal Louisiana's communities and transformed the region—once a place of refuge for displaced Native peoples and ethnic minorities—into what I will argue is an energy sacrifice zone that risks setting the stage for future disasters (Figure 0.1).

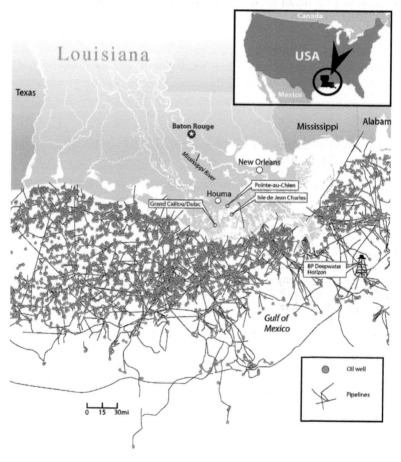

FIGURE 0.1 Map of the three tribal communities and pipelines off the coast of Louisiana.

There are approximately 25,000 miles of pipelines and 3,500 offshore production facilities in the central and western Gulf of Mexico's federal waters, three-quarters of which are off the coast of Louisiana (Freudenburg and Gramling, 2011: 171). The three tribal communities discussed in this book are located amidst this sacrifice zone. Map by Donna Gayer, Artasaverb.

While the focus here is predominantly on three tribal communities in Louisiana, this does not preclude the fact that many diverse communities around the world are being affected by the impacts of climate change, habitual disasters, and the extraction and production of fossil fuels. The focal choice is not meant to discount the effects on others but rather to illustrate how the lived experiences and knowledge that emerge from the communities included here can provide guidance for what is coming down the pipeline, and in some cases already unfolding, for other populations and places, both rural and urban.

The evidence provided is a call to action to learn from the individuals and communities living the reality of rapid environmental change every day. Their insight—based on long-term observation, monitoring, and wisdom that has been passed down over generations—can inform effective, responsible, and needed change for the sake of our future, all of our relatives, and our shared home.

Notes

1 Conversation with Chief Albert Naquin in Isle de Jean Charles, Louisiana, January 4, 2012.
2 Truehill (1978).
3 Basso (1996).
4 Couvillion et al. (2011); CPRA (2012).
5 Barry (1997); Turner (1997); Penland et al. (2000).
6 Conversation with Kerry St. Pé in Thibodaux, Louisiana, July 27, 2012.
7 The word "bayou" originates from the Choctaw word *bayuk*. Plural "bayous" refers to all of the individual bayous flowing in the Bayou region of southeast Louisiana. "Down" the bayou means south; "up" the bayou indicates north.
8 Tribal members live in these settlements. However, there are a number of tribal members that have relocated from these settlements due to disasters, habitual flooding, and other reasons, although many still live nearby and throughout the region.
9 "Atakapa" is also spelled "Attakapa."
10 Parish governments, similar to a county, exercise a variety of different functions, such as maintaining many water works, roads, health units, and hospitals; promoting economic development; regulating business activities; and overseeing many state and federal programs in the parish (Police Jury Association of Louisiana, 2014). Coastal parishes in Louisiana also have levee districts that engage in flood control projects at the parish level, such as the Terrebonne Levee and Conservation District.
11 The original name for "Cajuns" is Acadian, but those who migrated to Louisiana are commonly referred to as "Cajuns."
12 All uses of climate change in this book refer to anthropogenic (human-caused) climate change.
13 Sense of place can be described as the "connection between people and the places they repetitively use, in which they dwell, in which their memories are made, and to which they ascribe a unique feeling" (Morgan et al., 2006: 706).
14 Oliver-Smith (1999); Button and Oliver-Smith (2008); Reed (2008).
15 Maskrey and Peacock (1997).

1

A CLIMATE OF CHANGE

Oh yeah, global warming and with global warming, rising sea levels. We have El Niño and La Niña in the Gulf. Have seen those effects . . . higher activity of hurricanes and stronger and damage goes further . . . that's all newer stuff coming out and it's because of climate change. Mother Earth is mad, she's angry. Look at what we're doing to her. We're polluting her air, damaging her grounds, polluting her waters, how do you want her to fix the problems when you're doing so much she can't keep up.

—Chief Shirell Parfait-Dardar, Grand Caillou/Dulac[1]

Earth's climate—typical weather trends over long periods of time—has changed numerous times due to natural events throughout history, shifting between long ice ages to periods of warming. However, since the start of the Industrial Revolution in the eighteenth century, the planet began to heat up, primarily due to human activities releasing immense amounts of greenhouse gases (carbon dioxide, methane) into the atmosphere through processes such as burning of fossil fuels (coal, oil, natural gas), deforestation (which releases the carbon stored in trees), industrial agriculture, and contemporary transportation. Accumulating in the atmosphere, the overabundance of greenhouse gas concentrations result in trapping heat around the planet, causing global temperatures to rise.

The abundant scientific evidence is clear. As early as 1896, the Swedish chemist Svante Arrhenius predicted that carbon dioxide (CO_2) emitted into the atmosphere from the burning of coal would eventually increase Earth's temperature. Decades of research followed. Nearly 100 years later, an unequivocal call to action was presented to the United States (US) Senate. In 1988, the National Aeronautics and Space Administration's (NASA) climatologist James Hansen testified before the US Senate about the relationship between global temperatures, global warming, and

the greenhouse effect. He demonstrated through research from NASA's Goddard Institute of Space Studies that human activities, predominantly the emissions of greenhouse gases, are primarily responsible for contemporary climate
change.

Sixteen of the last seventeen years have been the warmest years on record for
the planet. Climate records show that average temperatures have been much
higher and risen faster in the last century than any time in the past 1,700 years.
Global annual average temperature has increased by more than 1.2°F (0.7°C)
between 1986 and 2016, relative to 1901–1960, and by 1.8°F (1.0°C) from
1901 to 2016. Natural climate variability has caused only a small fraction of the
change. The increase in average global temperatures relative to pre-industrial
times could be limited to 3.6°F (2°C) if there is a drastic reduction in greenhouse
gases emitted into the atmosphere. However, following the high-emissions path
we are on, the temperature increase could reach 9°F (5°C) or more by the end of
the twenty-first century.[2]

Climate Change and Sea Level Rise

Global average sea levels have risen 7–8 inches (16–21 cm) since 1900. About three
of those inches (7 cm) have occurred just since 1993. Global mean sea level is very
likely to rise 1.0–4.3 feet (30–130 cm) by 2100, relative to the year 2000. For high-
emissions scenarios, which is the current trajectory, and based on emerging science
on the Antarctic ice sheet stability, a global mean sea level rise exceeding 8 feet (2.4 m)
by 2100 is physically possible.[3]

Rapid, multi-meter sea level rise is now projected to begin much sooner than
previously assumed, with continued rates of CO_2 emissions locking in unavoidable
consequences that we still do not fully understand.[4] In the United States, sea level
rise, more frequent and intense hurricanes, and associated impacts are particularly
consequential for coastal residents. Relative sea level rise in the US Northeast and
western Gulf of Mexico is likely to be greater than the global average. Sea level rise
will increase the frequency and extent of extreme flooding associated with coastal
storms. The frequency of the most intense of these storms is projected to increase.[5]
Further, as the climate continues to warm, hurricane-induced rainfall rates are projected to increase, resulting in increased flooding for states along the Gulf Coast,
such as Louisiana.[6]

Climate Change, Disasters, and Displacement:
An Escalating Reality

Already, in some extreme circumstances, coastal territories around the world are
becoming unviable to maintain livelihoods and settlements due to increasing flooding, coastal erosion, sea level rise, and/or melting permafrost.[7] As early as 1990,
the Intergovernmental Panel on Climate Change (IPCC)—the international body
that assesses the science on climate change—identified climate-induced human

migration as a critical issue and potentially the greatest climate change impact on society.[8] After decades of evidence, the Fifth IPCC Report, released in 2014, affirmed that anthropogenic climate change is already causing, and will continue to cause, the displacement of entire communities. Climate-driven displacement varies from temporary (e.g., flooding) to permanent (e.g., sea level rise). Yet, with increasing climate risks, displacement is becoming more likely to involve permanent migration. In particular, three main climate impacts driving migration include sea level rise, drinking water availability, and extreme weather events.[9]

In 2016, more than 31 million people in 125 countries and territories were displaced by disasters. On average, 26 million people are displaced by disasters such as floods every year. That equates to one person displaced every second.[10] These numbers will surely rise as extreme weather events are predicted to become more frequent and severe with a changing climate.[11] Not every individual extreme weather event can be directly connected to climate change. However, there are distinct trends in the intensity, frequency, and duration of extreme events, which, over the long-term, are influenced by a changing climate. The issue of displacement becomes all the more heightened as tipping points and thresholds in the climate system are crossed, with direct implications on the intensity, frequency, and duration of extreme weather events.

Projections vary from 25 million to 1 billion environmental migrants worldwide by 2050.[12] A quantified account of people displaced by climate change impacts is the often-cited figure from the economist Nicholas Stern's *Review on the Economics of Climate Change* (2006), which estimated 150–200 million people would be displaced due to climate change by 2050. However, in a speech given in 2008, Stern acknowledged, "We badly underestimated the degree of damages and the risks of climate change. All of the links in the chain are on average worse than we thought a couple of years ago."[13] At the United Nations Framework Convention on Climate Change's Conference (UNFCCC) of Parties in Poznan, Poland that same year, the Deputy High Commissioner for Refugees, Craig Johnstone, announced that even by the most conservative predictions, as many as 250 million people would be displaced by 2050 by climate-related impacts such as decreased water supply and scarce-resource-driven conflict.[14]

Tens of millions of people in coastal areas, on river deltas, and on islands are at particular risk of displacement from sea level rise and higher frequency of storms and floods. In Bangladesh alone, potentially 25 million people are at risk of displacement by sea level rise over the next forty years.[15] In the United States, one recent study concluded that a sea level rise of 2.95 feet (0.9 meters) by 2100 would put 4.2 million people at risk of inundation, whereas a rise of 5.9 feet (1.8 meters) puts 13.1 million people at risk, when taking into account projections of future population growth and migration.[16] This is especially alarming considering that with increased emissions, temperature rise, and melting glaciers, worst-case scenarios are already being exceeded.[17]

However, the numbers only tell a fraction of the story. The estimates put forth fail to illustrate the dynamic interaction of climate change with other factors and the ways in which climate change can act as a threat multiplier. For example, over 10 million Bangladeshis have moved to neighboring Indian states over the past two decades due to population growth and land scarcity, exacerbated by perpetual floods and droughts affecting livelihoods, as well as coastal land loss due to inundation from sea level rise.[18] Host communities face increased economic and natural resource competition as displaced households from flooded areas migrate to these regions, resulting in indirect effects from climate impacts occurring far from their homes. The numbers also blur the many distinctions among displaced communities and among types of migration and displacement patterns that may ensue.[19]

There is an ongoing debate among policy and decision makers, practitioners, and researchers about the perception of viewing human movement as a positive or negative adaptation strategy to climate change. According to the International Organization for Migration (IOM, 2009), "Migration often seems to be misperceived as a failure to adapt to a changing environment. Instead, migration can also be an adaptation strategy to climate and environmental change and is an essential component of the socio-environmental interactions that needs to be managed." However, the intersecting stressors to at-risk populations, such as poverty and health, hinder their ability to adapt, and successful relocation is not always possible. The most at-risk people often find themselves unable to leave because they do not have the resources to overcome the barriers to migration, placing their lives directly in danger and infringing on their human right to a healthy and safe environment.[20] Further, many frontline communities are place-based; migrating away from their homelands could result in loss of culture, livelihoods, health effects, and forced assimilation, among other impacts.

Indigenous[21] Communities on the Frontlines

Traditional knowledges,[22] long-term observations, and cultural practices have guided Indigenous peoples' adaptation to environmental change for millennia. Further, traditional knowledges are increasingly recognized as necessary and valuable to inform and guide climate adaptation,[23] with what Native scholar Dan Wildcat (2009, 2013) calls "indigenuity." For instance, anthropologist Kristina Peterson (2014: 4) noted that the Indigenous peoples who have spent centuries dwelling in coastal Louisiana living off the land and waters "are seeing the changes that are happening, whereas much of the rest of Western society is not paying close attention."

Humans live in an environment shaped by natural processes and human actions, subject to continuous change and alteration. At the same time, the pace at which ecological change is now occurring is often outpacing traditional adaptive strategies and threating traditional knowledges and understandings of the relationship between ecological events and traditional and subsistence practices. The current context of changing climatic conditions makes it more challenging to predict the weather, and traditional calendars for the timing of natural events no longer function.[24]

After centuries of colonial history and policies reducing adaptation options, a number of Indigenous communities, cultures, traditions, practices, and ways of life, which are dependent on intact ecosystems, are at high risk from climate change impacts and ecological dispossession.[25] Further, many Indigenous communities have been forced onto lands exposed to weather extremes and vulnerable conditions. Restricted to reservation boundaries and forced into settlements, as ecosystems shift and migrate, people can no longer migrate with them, limiting their ability to adapt to ecological change.[26] They are also among the groups that have contributed the least to greenhouse gas emissions that drive climate change. In fact, many Indigenous communities at risk of climate change impacts are the same ones that have already been—and continue to be—sacrificed by the fossil fuel extractive energy industry.

The Climate Change–Fossil Fuel Extraction Connection

Accelerating rates of fossil fuel extraction and production, such as a projected increase of natural gas production in the United States by 55 percent by 2040,[27] will lock in enough carbon emissions to bust through agreed climate goals,[28] signed by 178 nations in the 2015 Paris Agreement, which established a global action plan to limit global warming to below 2°C. This is the same agreement that the Donald Trump Administration withdrew the United States from in June 2017.

In 2016, the global mean atmospheric CO_2 concentration—largely from burning fossil fuels—exceeded 400 parts per million for the first time in the last 800,000 years, based on evidence from ice core records. Between 2015 and 2016, carbon dioxide, methane, and nitrous oxide—the dominant greehouse gases—reached new records for the largest annual increase of emissions into the atmosphere observed in the 58-year measurement record.[29] The present-day greenhouse gas emissions rate of nearly 10 gigatonnes of carbon, or 10 billion metric tons, per year suggests that in the last 50 million years there is no climate equivalent for the twenty-first century.[30] We are in uncharted territory.

Additionally, climate change is, at times, being used by certain entities to further geopolitical goals of fossil fuel extraction. In the Arctic, declining sea ice has enabled the opening of passageways for navigable transportation to reach what is estimated to be nearly one-quarter of the world's undiscovered oil and natural gas deposits, mostly offshore.[31] In some coastal areas climate change can be used by oil and gas corporations to shift the blame for extreme land loss, escaping accountability for restoration caused by oil- and gas-related activities. Kathy Mulvey, Seth Shulman, and others with the Union of Concerned Scientists (2015: 1)—a non-profit science advocacy organization—have raised concerns that for nearly three decades major fossil fuel companies have "knowingly worked to distort climate science findings, deceive the public, and block policies designed to hasten our needed transition to a clean energy economy. Their tactics have included collusion, the use of front groups to hide companies' influence and avoid accountability, and the secret funding of purportedly independent scientists."

The American Petroleum Institute (API)—the largest US trade association for the oil and natural gas industry and the industry's principal lobbyist—is among the groups accused of public deception about climate change. Mulvey et al. (2015: 9) further raised the issue that to thwart approval of the 1997 Kyoto Protocol—the international agreement for participating countries to commit to binding emissions reductions—and other climate-related policies in the United States, the API Global Climate Science Communications Team, which "consisted of representatives from the fossil fuel industry, trade associations, and public relations firms . . . mapped out a multifaceted deception strategy for the fossil fuel industry that continues to this day—outlining plans to reach the media, the public, and policy makers with a message emphasizing 'uncertainties' in climate science."

Further, NASA scientist James Hansen, who first called for climate action in his 1988 testimony to the US Senate, as discussed earlier, documented increased political interference with scientific testimony to Congress since the early 1980s, including on testimony related to climate change.[32] While Hansen raised the issue during the George W. Bush Republican Administration, which strategically blocked climate legislation and distorted climate science findings,[33] the interference occurred over time under both Republican and Democratic oversight. The current Trump Administration's recent actions to dismantle progress made on climate adaptation and mitigation, and further public deception – such as removing the US Environmental Protection Agency's climate science website from public view, appointing climate skeptics or those who do not perceive the need for urgent climate action to the top federal agency positions (leading to intensified suppression of scientific findings),[34] reassigning federal agency scientists and climate policy officials working on imminent climate-induced threats to Native communities to obscure, unrelated jobs,[35] removing the United States from the Paris Agreement, and prioritizing permitting oil, gas, and coal projects as a matter of national security,[36] amidst dozens of other actions—is not entirely new, but rather a ramped up, steroidal hit to what has been unfolding for several decades.

Symptoms of Deeper Pathologies of Power

Climate and other environmental changes are not an aberration but rather, what medical anthropologist Paul Farmer (2003: 7) described as, "symptoms of deeper pathologies of power," reflected in the prevailing economic, political, and social systems, which play out in people's lives through forms of historically and economically driven structural violence. A single identifiable actor does not cause structural violence. Rather, structures or systems cause harm through the "social machinery of oppression,"[37] which also masks such violence.

The converse relationship between those responsible for climate change and those most affected highlights the power dynamic between the people who control policy decisions and access to resources and the people whose power is constrained by modern economic and political processes. This is akin to what Native environmentalist and activist Winona LaDuke (2013) described as "predator economics," in which the greatest negative impacts of the economic system are wrought on places where people have the least resources and legal knowledge to resist.

Often, climate change effects and other environmental stressors impact people who have already been pushed to the fringes—geographically, economically, politically, and socially. Such effects can act as a tipping point to the changes to which people have already been adapting. Tribal and other communities in southeast coastal Louisiana are living this reality every day.

Notes

1 Conversation with Chief Shirell Parfait-Dardar in Chauvin, Louisiana, May 16, 2012.
2 USGCRP (2017: 15).
3 Ibid.
4 Hansen et al. (2015).
5 USGCRP (2017).
6 Frankson and Kunkel (2017).
7 Bronen et al. (2018).
8 IPCC (1990).
9 Adamo and de Sherbinin (2009).
10 NRC (2016).
11 IPCC (2014); Melillo et al. (2014).
12 IOM (2017).
13 Fortson (2008).
14 Sunjic (2008).
15 Displacement Solutions and YPSA (2014).
16 Hauer et al. (2016).
17 Hansen et al. (2015).
18 IPCC (2014).
19 Marino (2015).
20 Office of the United Nations High Commissioner for Human Rights (2009); Gemenne (2010).
21 I use the term "Indigenous" here as a broader term that encompasses (Native American, Alaska Native, Native Hawaiian, etc.). In the United States, the use of American Indian or Native American is often associated only with federally recognized tribes, whereas the broader use includes state-recognized and non-federally recognized tribes and those who self-identify as Indigenous.
22 "Knowledges" is used in the plural form here, following the recommended use in the Guidelines for Considering Traditional Knowledges in Climate Change Initiatives. The Guidelines define traditional knowledges as "Indigenous communities' ways of knowing that both guide and result from their community members' close relationships with and responsibilities toward the landscapes, waterscapes, plants, and animals that are vital to the flourishing of Indigenous cultures" (Climate and Traditional Knowledge Workgroup/ CTKW, 2014: 7).
23 CTKW (2014); IPCC (2014).
24 Blanchard (2015).
25 Maynard (2002, 2014); Whyte (2013).
26 Lynn et al. (2013); Whyte (2014).
27 The projected increase of natural gas production in the United States is in large part a result of the use of hydraulic fracturing and horizontal drilling to access previously inaccessible gas formations (US EIA, 2016).
28 Stockman (2016).
29 Blunden and Arndt (2017).
30 USGCRP (2017: 31).
31 USGS (2008); Arctic Council (2009).

32 Hansen (2007).

33 For more details, see Wood (2013).

34 For example, the US Department of the Interior was reported to have recently suppressed the conclusions of a sea level rise study (for more information, see Hanley, 2017).

35 Clement (2017).

36 For example, see the Trump Administration's priority work list for the Bureau of Land Management (Streater, 2017).

37 Farmer (2004: 307); also Farmer (2003).

2

ENTRÉE INTO COASTAL LOUISIANA

Seeing the Unexpected

I left our small dwelling, known locally as a camp, in Pointe-au-Chien at sunrise. The heat was already emanating from the road on the late August morning in 2012. The air was still. Almost too still. I ran the mile down to the marina at the south end of Bayou Pointe-au-Chien on the Terrebonne Parish side, across from the Pointe-au-Chien Indian Tribal Community's side of the bayou in Lafourche Parish. The sun rose over the water and enveloped the marsh grass in yellows and reds. Down at the marina, I saw a few people fishing from the dock. Four horses grazed on the ridge to the east, as the rising sun reflected off the poles of the shrimp land nets. Running north back up the bayou, water flowed slowly next to me, as did a local fisherman and his son returning home from a night of shrimping, pulling up netting in the back of their boat. A few pelicans flew across the bayou to perch in dying trees along the water's edge. Several trucks passed, parking along the road to establish their fishing spot for the day.

Crossing the bridge in front of the sign—Pointe-au-Chien Indian Tribe—I looped around to the other side of Bayou Pointe-au-Chien. I ran about one and a half miles south until I hit water. I turned around at the last trailer, located next to stilts jutting into the air. The house that had once rested on the stilts and attached platform was destroyed during a recent hurricane. A friend was nearby hauling crab traps out of the water in preparation for the approaching storm. He told me he had not caught any crabs anyways with everything that was going on in the water, referring to the diminished catch since the 2010 British Petroleum (BP) Deepwater Horizon Oil Disaster. I jogged by the one small business left in the community where local fishers sold their shrimp; several young men were standing outside sorting the shrimp bought that morning. I looped back across the bridge and soon turned down the Island Road, connecting Isle de Jean Charles to Pointe-au-Chien and the mainland. I came across Theodore, from Isle de Jean Charles, who was driving to a store to get a newspaper. Eighteen miles round trip to get a paper, as delivery service had stopped to the Island after the road was damaged during Hurricanes Gustav and Ike in 2008; the service did not resume once the road was fixed.

I ran slowly down the Island, listening to birds chirping nearby and the faint sound of a radio. I passed a shed surrounded by a décor of dozens of cans tied up together in nets with cardboard boxes strewn about on the ground. A subtle breeze momentarily cooled me as I looked beyond the shed and saw the cross planted in the ground, marking the cemetery. Between the scattered houses remained a few standing posts, remnants of a trailer that was bulldozed a few months prior but not yet removed, and a number painted on the road indicating an address, but there was no house left standing.

I arrived at a house belonging to Chris, a life-long Island resident. He was sitting underneath his elevated house with a sack of oysters in front of him on the table. He had a glove on one hand holding an oyster that spanned the length of his hand, and held a knife in his other hand. He opened up the oyster and handed it to me on the half-shell. The saltwater exploded in my mouth as I looked up at the lightning streaking across the darkening sky. Sheets of rain came down in the distance. Cutting my visit short, I knew I didn't have much time to get back across the Island Road. The lightning struck again.

As I ran back across, the sun's rays peaked through the clouds and shined out over the water, across the eroding marsh and the remaining skeletal trees. I was transfixed—and suddenly very aware of being surrounded by water. Despite the imminent storm, two men got out of a truck to cast their fishing rods along the side of the road. I watched a few herons spread their wings and take flight over the marsh grass, which quickly disappeared into water.

Later that night with the storm approaching, I walked alongside Bayou Pointe-au-Chien. Land nets placed in the water to catch shrimp cast out a line of lights. A skiff with a man and young boy slowly passed in the bayou. The reflection of light from the boat shimmered across the dead trees. I saw the ripples in the water alongside me where a fish had just jumped, heard the frogs croaking in the marsh grass just over on the other side of the levee, and tasted the slight hint of salt in the air. What at first seemed so still suddenly became full of sounds, smells, and tastes, of life, livelihoods, and culture.

My Entrée

My first visit to Louisiana's bayou region was in September 2009, four years after Hurricanes Katrina and Rita and one year after Hurricanes Gustav and Ike tore through the area. Spending the late night hours on a shrimp boat, through the invitation of a trusted local community organizer and researcher, I watched the porpoises below the water and stars above, the disappearing marsh, and remnants of houses left behind from recent hurricanes. I fell in love with the quiet beauty of the natural environment and generosity and openness of the people. In that moment, I started relationships that have turned from one of stranger to family.

I returned briefly in June and July 2010, weeks after the explosion of the Macondo Well in the Gulf of Mexico and the official beginning of the BP Deepwater Horizon Oil Disaster. Subsequently, I conducted full-time fieldwork from October 2011 (a month after flooding from Tropical Storm Lee) until September 2012 (right after Hurricane Isaac) for my doctorate in anthropology. During this time, I engaged primarily with three tribes in the region—the Isle de Jean Charles, Grand Caillou/Dulac, and Pointe-au-Chien Indian Tribes. I returned a couple months later in December 2012, witnessing the continued, ongoing effects of Hurricane Isaac and the BP Deepwater Horizon Oil Disaster. I continue to visit and spend time with the communities and to engage with tribal leaders and organizers.

Often asked how a young woman from California who had worked in various places around the world ended up in coastal Louisiana, I never really knew what to say, as I hadn't quite figured it out yet myself. I was drawn to the fundamental injustices of displacement at the intersection of climate–disasters–energy, but why here in particular? The answer slowly revealed itself to me one evening in the early spring of 2012. I walked with a friend behind the house of two elders on Isle de Jean Charles. We stood up on the levee, which had been built by the local parish in an attempt to stop the flooding during high tide. We looked at the sun hanging just above the water. The reds, oranges, and yellows deepened. My friend turned to me and asked, "You ever get that feeling like you're meant to be right where you are?"

Placing Myself

One of the biggest challenges of anthropological fieldwork is figuring out how to place one's self. I first situated myself physically in Houma, Louisiana, which is like the palm of the hand, with the bayous as the fingers stretching out from it. Houma is about 20 miles north of the three tribal communities with roads connecting to each community. Once I felt I was not over-stepping boundaries, I found a camp to live in with my partner and dog toward the southern end of Pointe-au-Chien on the Terrebonne Parish side. Living in the community, I accompanied people on their daily tasks and worked alongside them, seeing the ways families interacted together and shared resources, and witnessing people who had relocated nearby still coming and going down their home bayou.

Living there as Hurricane Isaac approached at the end of August 2012, I saw how people tracked the storm, made decisions to leave or stay and where they would go; how they worked together to get boats moved, crab traps collected, supplies divvied out; and how their way of being together, of celebrating, of adapting, continued. In the aftermath of the hurricane, I handed out food and water supplies in Pointe-aux-Chenes— approximately 10 miles north of the Pointe-au-Chien Indian Tribal Community.[1] As I stood in the middle of the road directing traffic and chatting with friends passing by, I realized that I was no longer in this abstract "field." I felt that I was somehow home, a testament to the openness and sense of community shared by local residents.

The Fine Line between Research and Extraction

During the course of my research, I watched and participated in interactions between local residents and other researchers, journalists, activists, filmmakers, and government and organizational representatives. I saw those who reached out with the best of intentions and fell in love with the people, culture, and place. I witnessed how others took advantage of people in a vulnerable situation.

Not wanting to be another source of extraction, I tried to acknowledge and remain aware of who is silenced in the research process, acknowledging my own role in the power relationship. To conduct responsible ethnographic work it is essential that we learn from our assumptions and, when we engage with participants, we genuinely listen. We have to be ready to see what we do not expect.[2]

When I started my fieldwork I assumed that local residents' perspectives about the causes of local environmental changes would be more balanced between the oil industry's dredging canals and the government's manipulation of waterways (e.g., the building of dams, levees, and other flood protection measures). The State of Louisiana has historically asserted that coastal land loss is due to sediment deprivation, and the government and media often publicize the extreme land loss in coastal Louisiana as predominantly caused by the developed waterways; thus, restoration needed to be based on sediment management.[3] However, this is only part of the story. The vast amount of land loss since the 1930s has largely been due to the impacts of changes in wetland hydrology from oil and gas corporations dredging channels and forming spoil banks,[4] which are excess surface materials left alongside a canal after dredging.

Local residents expressed to me their frustration over the misunderstanding of the causes of land loss. Almost everyone I spoke with identified the oil industry dredging canals for pipelines and industry-related navigation as the biggest cause of local environmental changes—land loss, coastal erosion, and saltwater intrusion. Some examples of what people expressed include:

> Oil companies got down there and started building canals everywhere, that's when the land started going away.
> *Josette, Isle de Jean Charles tribal member, relocated 20 miles northeast*[5]

> Like you see right here, that's a pipeline and had they not cut into this, that would be land right now. Everywhere you look there's pipeline. You can go ride in the water, there's pipelines.
> *Regina, Isle de Jean Charles resident*[6]

Coastal land loss is the result of complex natural and human interactions. It is thus difficult to isolate any one activity as the sole cause of specific land loss; often, local residents would point to overlapping factors. For example, when I asked Gabrielle, who relocated as a child with her family from Isle de Jean Charles to Houma after their trailer flooded during a hurricane and the mold took over, about the biggest cause of land loss, she described a multitude of converging issues,

> A lot of it has been with gas and oil exploration, the cuts, the cutting of canals and things, the saltwater. The saltwater is what killed everything . . . the water flowing a certain way is what built the land. . . . Because land gets rebuilt a certain way and when it's been built like that for thousands of years and all

of a sudden you decided to come in and cut it and change the flow of that, you're going to have losses and the losses now are outweighing the gains.[7]

When I asked local residents about land loss, some would talk about the construction of dams and levees and diverted waterways, or other issues such as invasive species (e.g., nutria), natural subsidence processes, and government regulations (e.g., hunters and trappers were no longer allowed to burn the marsh, a community land management practice used to rejuvenate the marsh). But when asked about the primary cause of environmental changes, first and foremost people talked about how dredging canals for oil and gas pipelines and navigation predominantly for oil and gas extraction caused the land loss and saltwater intrusion.

Struggling with how to tell a different version of a story than what was typically publicized, I asked my partner, a civil engineer, what he thought. He had always been told in professional circles and understood the land loss in the region as the result of the manipulation of waterways by building dams and levees. But after seeing the landscape covered with oil refineries and the waterways filled with oil rigs and channels for pipelines, he realized there was more to the story: "It's the people living with pipelines in their backyard . . . this is what people should know."

In paying attention to my own placement, subjectivity, and research methods, I hope to share what shaped my understanding, theoretical insight, and conscientization[8] about people's lived experiences of rapid environmental change. Focusing on their stories and perspectives, I am producing a subjective account of reality, acknowledging my own role and views in interpreting their experiences.

As I became involved in the internal community politics and befriended local residents, I understood my responsibility as a researcher more clearly and ethical choices that I needed to make. I aimed to try to ensure that what I produce does not have unintended negative consequences on activities the communities are pursuing but rather supports their efforts—that I am not just extracting knowledge *from* them but rather working *with* them. As one way to honor their voices and wisdom, most chapters of this book start with a quote from different digital stories I created with tribal members, explained below, which is then accompanied by the full story on a following page.

Research Methods for an Advocacy Anthropology

Different visitors to the communities often asked me if I was an activist or advocate. I would correct them by saying I am an anthropologist. But I now realize it is not necessary to have to choose between the two. An advocacy anthropology that is shaped to reach public audiences, decision-makers, and non-anthropologists works to decolonize the research process and is built on the notion that research is about knowledge that is most useful for people to problem-solve and translate into action.[9] Advocacy goals shaped my research practice and choice of methodologies, as well as my community work such as helping with community meetings, researching for the Tribes' federal recognition process, organizing documentation for the

Tribes' input into national-level government reports, and speaking to audiences around the country, including Congressional members and staff, about the urgent needs of the Tribes and ways that they are addressing the challenges.[10] Part of my advocacy effort includes writing and publishing in various outlets, including this book, in an attempt to create awareness among diverse audiences.[11]

To initiate the research, I started with the Tribes' leaders, as is culturally appropriate, as well as the Tribal Councils when asked by the leaders, and asked their permission to conduct research with the Tribes.[12] I created a "statement of purpose" that included my research objectives, methodology, and a set of guiding principles I would follow while conducting the work.[13] The following methods enabled me to better understand the long history of displacement and environmental degradation the three Tribes experienced and drew out people's reflections about the environmental changes in ways that might not have otherwise been revealed.

Participant Action and Reflection

The main research method I employed was participant action, taking part in everyday community activities, gatherings, and social interactions, followed by reflecting on what unfolded. Through my placement and active participation, I gained insight into local residents' lived experiences through their culturally specific context and ways of knowing.[14] I participated in local ways of food sharing, such as bringing shrimp to an elderly resident or coming home to a pile of boiled crabs at my doorstep. Attending Tribal Council and community meetings helped me understand the most important current issues for the Tribes, such as obtaining federal recognition and the BP Deepwater Horizon Oil Disaster settlement claims. Out on a boat shrimping late at night or hanging out with shrimpers at the land net at the end of Pointe-au-Chien, I watched how informal trading occurred with customers and friends, saw firsthand the shrimpers' diminished catch, and heard stories about land loss and other environmental changes being experienced, local knowledge of the waterways, and why it was so important to many local residents to stay in place.[15]

Intentional Conversations

Instead of a structured interview guide, I had intentional conversations with tribal members—both those still living in the geographic communities and those who had relocated nearby—during which I wove specific topics into one-on-one conversations, with the goal of providing culturally appropriate space for dialogue. The conversations, mostly held at people's houses, were aimed to be based on the way people were already used to interacting and engaging, talking over many hours and not confined to a set of questions and schedule. This type of engagement helped me better understand what was most important to them, pointing to ideas and issues I would not have thought to ask about; the seemingly tangential topics often ended up being the heart of someone's story.

Story Circles

I facilitated several story circles with each community. Four to six people participated in each one, using the mode of communication that people were accustomed—being together and sharing stories (Figure 2.1). For each story circle, we spent several hours gathered together around a table in a participant's house, sharing food and conversation. I posed some broad questions to start the conversation, with the participants then taking the discussion in their own direction. Storytelling can be a healing process, but it can also be taxing to dredge up certain memories and enhance stress to which I needed to be aware of and remain sensitive to throughout the process.

The story circles helped me learn not just *what* is important but also *how* local residents remember, such as events and memories are not marked by time but rather by hurricanes. I learned about important gathering places, local place names, and places that no longer existed as the land washed away and infrastructure flooded and ripped apart during storms. The participants talked about what their lives were like growing up, about spiritual and religious beliefs, and about the experience of segregation and being punished for speaking French in school, which older adults and elders from the communities grew up predominantly speaking.

I more readily understood the communities' interactions with local government officials and agencies and the politics of levee placements, along with the causes and consequences of being mostly left out of government-led restoration and flood mitigation plans. I learned about the first oil and gas corporations that came to the area, the environmental changes the residents started noticing after canals were dug for

FIGURE 2.1 Story circle, Isle de Jean Charles.

Source: Photo by author, 2012.

oil and gas pipelines and navigation, and why people often had no other choice but to work for the oil corporations or related industries after losing their fishing-based livelihood and with limited access to other economic or job opportunities. They voiced their perceptions about their community's future and options for adaptation.

Digital Storytelling

I also created digital stories with six tribal members. Digital stories are short narratives created in video format that bring together participants' words, voices, and images. The idea was for the stories to be useful to the communities for their advocacy efforts; for example, a tribal leader used her story to apply for a grant. With permission from the storytellers, the stories have been shown at public events, film festivals, and professional conferences and agency forums to bring attention to the issues that the Tribes and other coastal Louisiana communities are experiencing.

Stories were written by the storyteller or told to me over several conversations. We went through an iterative editing process, which maintained the storyteller's own words and ideas. One major challenge with this method was that it imposed an unfamiliar way of storytelling on the participants, by setting a boundary on the length of the story. I adapted the techniques to try to fit what worked best for the storytellers. For example, I sat and took notes as Marlene, Grand Caillou/Dulac Tribal leader and elder, who had relocated about 20 miles northeast from Dulac to Bourg, spent hours telling me her story while cooking jambalaya. We then iterated about her final recorded story.

The process of creating the stories revealed new ways of seeing and experiencing the storyteller's world. Together with the storyteller, we looked through pictures of family gatherings, boat blessings, flooding, and post-storm debris, and drove around the area to capture the images they wanted to include. Storytelling shapes our experiences and how we learn in an engaged process.[16] The various senses people shared through the process—visual images, sights, smells, tastes—provided deeper insight into their relationships with the land, water, and their community, showing the interconnectedness between their subjectivities, environment, and experiences. I learned about how they perceived and understood the local environmental and social changes and what the changes have meant for their communities, families, and personal lives, experiencing a deeper way of knowing and understanding than would have been possible through passing conversation alone. The digital stories created provide a living testimony of the storytellers' lives, cultures, challenges, and resolve.

Cross-Community Conversations

Along with my local mentor, I facilitated conversations with communities in other regions to bring people together over virtual communication pathways across cultural and geographic boundaries to share and learn from each other's experiences.

The conversations offered ways to learn about similar environmental changes and obstacles experienced, the processes they were going through to address the adverse challenges, and how decisions were made. The conversations built and expanded upon work already underway by my local mentor and a team of local researchers and community organizers to bring communities together over common resource concerns.

Names

I originally intended to use pseudonyms for everyone who participated in the research. However, I soon realized that disguising the names of the tribal leaders would be unrealistic and potentially disrespectful. The same holds true for people I spoke with who hold public positions. For those who created digital stories, some of which have been shown publicly, it would also be difficult to hide the storytellers' identities. Therefore, I have established a mixed-pseudonym system, with special consent given by people whose real names are used. For other participants, I have created pseudonyms, following traditional anthropological protocol.

The Concept of Tribal Community in This Research

I use the term "community" here as a way to define a relationship between people shaped by a shared heritage. This does not mean that the people within these defined communities are homogenous; the individuals have different opinions and points of view, as well as their own individual experiences, memories, and knowledge.

There are residents within each of the geographic areas, particularly in the Grand/ Caillou Dulac area, who are members of another Tribe and residents from other diverse descents. Therefore, I do not claim this research to represent a study of any entire geographic area or to represent the perspectives of all members of any of the Tribes.

★ ★ ★

The following chapters tell a story about the socio-historical and current drivers of co-occurring adverse events and habitual disasters, accompanied by the risks of displacement that threaten the cultural sovereignty and social fabric of three communities that have dwelled in place for centuries. This is a story about the layers of disasters and vulnerabilities, but more so about what these communities, which are often seen in a dominant capitalist system as the "targets of least resistance,"[17] are actively doing to adapt, challenge, and resist structural violence, marginalization, and injustice within a rapidly changing environment.

Notes

1 The Pointe-au-Chien Indian Tribe and the geographic location of the tribal community is spelled "Pointe-au-Chien." The area approximately 10 miles north of the tribal community, still along Bayou Pointe-au-Chien, and where many tribal members now live, is spelled "Pointe-aux-Chenes" and often referred to by Pointe-au-Chien tribal members as "upper Pointe-aux-Chenes."

2 Bourgois and Scheper-Hughes (2004: 318).
3 For example, US Department of the Interior (1994).
4 Turner (1997).
5 Conversation with Josette in Grand Bois, Louisiana, March 9, 2012.
6 Conversation with Regina in Isle de Jean Charles, Louisiana, June 1, 2012.
7 Conversation with Gabrielle in Houma, Louisiana, July 29, 2012.
8 Freire (1970).
9 Freire (1970); Park (1997); Smith (2004).
10 See for instance, EESI (2015).
11 Some portions of the material in this book have appeared in journal articles and book chapters I authored or co-authored; see, for example, Maldonado (2014a, 2014b, 2016a, 2016b, 2017); Maldonado et al. (2013, 2015); Maldonado and Peterson (2018); Bronen et al. (2018).
12 One acknowledged shortcoming of this research is that I might not have encountered as great an array of differences in opinion as actually exist in the communities. As people paid attention to who was doing what within specific spaces, they were often aware of whom I was already interacting with, which could have closed me off from certain segments of the population. Some people also had research fatigue from talking with journalists, with other researchers, and at public and government forums. And while I spent a lot of time interacting with youth from the communities, my conversations with people under the age of 18 are not included here.
13 Bethel et al. (2011); Peterson (2011).
14 Pink (2009).
15 I frame this method as "participant action and reflection" instead of the traditional anthropological "participant observation" because much of the activities I was involved in I was actively doing or facilitating, and not often focused on the observational aspect in the moment. I aimed to take time once a week to reflect back on the week's activities and engagement, and consider what I had learned and experienced.
16 Freire (1970).
17 Oliver-Smith (2009).

3

CO-OCCURRING AND
ACCUMULATING DISASTERS

Our land and trees are dying, becoming more saturated with saltwater brought
by hurricanes. Just with wind now there is high water that has to be pumped
out. . . . The plants are losing roots and dying. There is nothing to hold the
land together. They have oil and gas under this earth and they're pumping it all
out. The Earth is purging itself.

—Marlene Foret, Grand Caillou/Dulac[1]

Chief Shirell (Grand Caillou/Dulac), who had relocated about 15 miles northeast
from Dulac to Chauvin, and I drove down the bayou in Dulac to Shrimper's
Row.[2] We stopped on the way to see where her family lived until Hurricane
Andrew hit in 1992 when she was 12 years old. She had not stepped foot on
the property since. She described how they used to play a mile back behind the
house in the woods. Now, there was a sparse layer of trees in front of us with
water directly behind. We walked through the overgrown weeds and shrubs on
what had been her family's driveway.

She gazed out into the woods and walked a few feet to touch the trunk of an oak
tree close to the road. A big smile swept across her face as she looked up and saw the
rope swing her family had tied to a branch years ago. Tears slowly rolled down her
cheeks. She said how they did not think we are worth saving. I asked who "they"
were: she explained—the government and corporations.

We made another loop and crossed a bridge onto Shrimper's Row. She said
over and over, "home." We continued south. Trailers and shotgun-style houses
disappeared behind us, as we now passed elevated vacation houses. She pointed
to where people from her Tribe had once lived. We drove by an open gate with a
sign "Southern Comfort," a subdivision with camps owned mostly by what locals

call "weekend warriors"—predominantly Anglo-Americans from the greater New Orleans area who travel south/down the bayous for weekend recreational escapes. We passed road names like Bud Light Court and Champagne Drive. Approximately 24 gated or fenced-off neighborhoods of fishing camps were constructed in southern Terrebonne, Lafourche, and Jefferson parishes from the late 1980s to mid-2000s.[3] The vast development expansion further degraded the wetlands with inlets cut for recreational boats to navigate. Local residents often found themselves competing for natural resources with recreational fishers and tourists. Chief Shirell pointed in all directions to show me where there used to be land, and where her tribal members had once lived; more people had been forced to move away after each hurricane and flood event, and as more land eroded away. And this is only a small part of the story.

This chapter provides an introduction to the three Tribes that are the focus of this book and some of the accumulating environmental disasters, political context, and accompanying effects.

★ ★ ★

Then and Now: A Tale of Life on the Bayou

By Marlene Foret, Grand Caillou/Dulac[4]

When I was 7 years old I started going to the Indian school in Dulac, located in the bayou country of Southern Louisiana. Only speaking French at home, I learned English when I entered school. I liked school, feeling like it was no different than the White school. Every year, my brothers and I were pulled out of school around October because Daddy was a shrimper and trapper. We would go to our camp a few bayous to the west, which we could only get to by boat.

Daddy leased property where he trapped muskrats, minks, coons, and otters. Daddy and Mama would skin the animals. My brothers and I would help put the skins out every day to dry, making sure to watch for weather in case we had to bring in the skins.

Daddy and my brothers would go out shrimping and sell what they caught to the dried shrimp platform. I stayed at the camp with Mama and we would fix the garden, filled with green beans, corn, potatoes, okra, cucumber, cantaloupe, tomatoes. We ate shrimp and crabs when they were in season. We didn't have a fridge, so we ate them fresh.

We also had a garden at our home in Dulac. Daddy's butter beans were so good; they were so big and still green in the pod. Our house was made of wood, and when I was little we had kerosene lamps and butane for the stove and heaters. If we were sick, Mama knew what herbs to use and she'd treat us with peppermint oil, fish oil, and purge us after each season.

Our neighbors were all aunts and cousins. Grandpa lived on the bayou side in a camp boat. There was a church near us and Superior Oil. Mama and Daddy told us they had people we stayed away from because they didn't like Indians. But it didn't bother me. I liked everyone just fine.

It was beautiful where we lived because all the trees were big and healthy. We'd sit under the trees and enjoy the fresh air the trees provided. We had a gumbo tree in the yard with roots that smelled like root beer that we'd use to make things, like filet. The tree died because of hurricane waters that came through. When hurricanes came, we would evacuate together on Daddy's boat. But when my youngest brother was born the boat was no longer big enough, so we'd evacuate to the school. Back then the water went down fast. Daddy and my brothers would go back to our house to wash the floors, clean the walls, and then come back for Mama and me after a day or two to bring us back home and we'd start over again.

When I got a bit older Daddy quit shrimping and got a job at the Indian school as a janitor and Mama as a cook. Every weekend we all got up at 6am to go to school to clean with Daddy. We scrubbed and waxed the floors and wiped the desks. I still remember the beautiful green tiles.

I quit school after the ninth grade when I was 17 to go work at the shrimp factory. I stood at a long table pointed in the middle with a trough that ran with water. I broke the heads off the shrimp and let the bodies run down the water.

In '79 I got my GED[5] and soon remarried and moved to Bourg, a couple bayous east and north of Dulac. I worked at the Grand Caillou/Dulac elementary and middle schools for years as a custodian, just like my daddy. But now the schools had been desegregated.

My home community is still an Indian community, but it's different now. There is no more trapping. Few people hunt. There are more effects from hurricanes than when I was growing up. We used to have two banks, a drug store, a clothing store, grocery stores, a fruit stand, a meat market, and shrimp factories. These all started closing down after Hurricane Andrew brought lots of water in '92.

Today, there is no more marshland, no more buffer zone because of land erosion. Our land and trees are dying, becoming more saturated with saltwater brought by hurricanes. Just with wind now there is high water that has to be pumped out. The water doesn't go down fast like it used to. It gets trapped behind the levees. You can't grow anything because of the water. The plants are losing roots and dying. There is nothing to hold the land together. They have oil and gas under this earth and they're pumping it all out. The Earth is purging itself.

Many people have relocated, spreading up the bayou. I would move back home in a heartbeat, but there is not enough land and our community is cut out of the Morganza levee system. We are trying to get federally recognized so we can maintain our community, our elders can be supported, and we can have our own schools back and our own education for our kids. I would give my life to see us federally recognized to show who we really are.

I am Indian. And proud of it.

FIGURE 3.1 A portion of Grand Caillou/Dulac. The patches of land in the back used to be connected as one big land mass.

Source: Photo by author, 2012.

FIGURE 3.2 Marlene Foret, standing where she grew up in Dulac. The house was lost to hurricane floodwaters.

Source: Photo by author, 2012.

★★★

The Three Tribes Today

The members of the Isle de Jean Charles, Grand/Caillou Dulac, and Pointe-au-Chien Indian Tribes are descendants of Biloxi, Chitimacha, Choctaw, Acolapissa, and Atakapa Indians who lived in the Mississippi Valley for millennia before European contact. Approximately 1.3 to 4 million Native people lived in and shaped the landscape of the southeast region of the present-day United States before European colonists arrived.[6] The southeast region included significant linguistic, social, and cultural diversity among the Native population. Groups typically lived in small villages comprised of families situated around a center area where political and ceremonial activities took place. Native people of the Mississippi Delta relocated seasonally, as necessitated by annual flooding. People adapted, gaining understanding and respect over generations of how to live in their wetland environment and where to go when the water level rose. The groups favored collective wisdom passed down over generations over knowledge derived from individual experience.[7]

Isle de Jean Charles Band of Biloxi-Chitimacha-Choctaw Tribe

Isle de Jean Charles (the Island) is located in Terrebonne Parish, between Bayou Terrebonne and Bayou Pointe-au-Chien (Figure 3.3). It is connected to the mainland by a narrow strip of asphalt that is quickly eroding away (Figure 3.4). The Island settlement, established in the 1830s, includes mostly Isle de Jean Charles Tribal members and is locally acknowledged as the Tribe's territory. The Island used to have two grocery stores, with one also serving as the church, school,

FIGURE 3.3 Isle de Jean Charles, Louisiana.

Source: Photo courtesy of Babs Bagwell, 2012.

and dance hall, but these were both lost to flooding. Now there is only a small privately owned marina at the south end of the Island.

As of 2015, the Island had lost approximately 98% of its 33,000 acres of landmass (Figure 3.5). There were 78 houses and approximately 325 people inhabiting the island in 2002. By 2017, with widespread, persistent flooding and loss of subsistence livelihoods, only about 25 houses and 70 people remained. These numbers represent only a portion of the 656-person Tribal membership. If immediate and inclusive restoration and flood protection actions are not taken, the Island could be gone before 2050.[8] Yet, based on cost-benefit analysis, the Island has been mostly omitted from coastal restoration efforts, which is elaborated further in Chapter 7.

FIGURE 3.4   The Island Road that connects Isle de Jean Charles to the mainland.

Source: Photo by author, 2012.

FIGURE 3.5   Isle de Jean Charles in 1963 (left) and 2008 (right).

Source: US Geological Survey.

Pointe-au-Chien Indian Tribe

The Pointe-au-Chien Indian Tribe is located in Lafourche Parish alongside Bayou Pointe-au-Chien with Oak Pointe Road running between the bayou and the houses (Figure 3.6 and 3.7). Bayou Pointe-au-Chien is the border between Lafourche and Terrebonne Parishes. Approximately one-third of Pointe-au-Chien's 688 members live close together in Pointe-au-Chien, about one-third live in other communities farther up the bayou, and another one-third reside out of the immediate area.[9] The tribal land included territory several miles south where some tribal residents once lived, but it is no longer habitable due to destructive storms, erosion, encroaching waters, and land loss.

Pointe-au-Chien once had four dried-shrimp factories, but these are now gone. As of 2015, there was still one small shrimp business in the community where shrimpers sell their catch (Figure 3.8 and 3.9). There are fishing camps owned by weekend warriors, a trailer camping park, and a privately owned marina along the southern end of Bayou Pointe-au-Chien on the Terrebonne Parish side.

Also on the Terrebonne Parish side, there is a Baptist church, another church a couple miles north, and a Catholic church about 10 miles farther north in Pointe-aux-Chenes—a predominantly Cajun community—near a small grocery store. Residents from Pointe-au-Chien and Isle de Jean Charles attend these churches and an elementary school near the Catholic Church.

FIGURE 3.6 Entrance to the Pointe-au-Chien Indian Tribal Community.

Source: Photo by author, 2012.

FIGURE 3.7 Bayou Pointe-au-Chien.

Source: Photo by author, 2011.

FIGURE 3.8 Donald Dardar, Co-Chairman, Pointe-au-Chien Indian Tribe, preparing his shrimp nets.

Source: Photo by author, 2012.

FIGURE 3.9 Shrimp boat, Pointe-au-Chien.

Source: Photo by author, 2012.

Grand Caillou/Dulac Band of Biloxi-Chitimacha-Choctaw Tribe

A portion of the approximately 1,200 members of the Grand Caillou/Dulac Tribe reside in Terrebonne Parish, along Shrimper's Row next to Bayou Grand Caillou, and other Grand Caillou and Dulac neighborhoods (Figure 3.10). These neighborhoods also include a larger Native population identified with the United Houma Nation, as well as Cajuns, Anglos of other European descent, and African-Americans, among others. People in the area tend to live clustered together by ethnicity and kinship groups.

Dulac is among the larger settlements associated with the coastal tribes. Dulac, including Grand Caillou/Dulac Tribal members and other residents of diverse descent, decreased from an estimated population of 2,458 people in 2000 to 1,463 people in 2010, with 39.4 percent American Indian and a median household income of $19,738.[10] The decrease in population size is largely attributed to Hurricanes Katrina and Rita in 2005, Hurricanes Gustav and Ike in 2008, and escalating flood insurance rates, discussed in more detail below. The Grand Caillou/Dulac area used to have grocery stores, shrimp factories, and a drug store, fruit stand, meat market, bank, and clothing store, but these started closing down after Hurricane Andrew hit in 1992, with more stores closing following Hurricanes Katrina and Rita in 2005. Now people living in the communities have to drive 30 minutes to get to a grocery store (Figure 3.11). There is a Catholic church in Dulac and an elementary and middle school in Grand Caillou. There is also a vast array of fishing camps owned by weekend warriors who use the area for recreational fishing on the weekends and holidays.

FIGURE 3.10 A few shrimp boats remain in Grand Caillou/Dulac.

Source: Photo by author, 2012.

FIGURE 3.11 Remnants of grocery store closed since Hurricanes Katrina and Rita in 2005, Grand Caillou/Dulac.

Source: Photo by author, 2012.

Federal Recognition

Grand Caillou/Dulac, Isle de Jean Charles, and Bayou Lafourche Tribes, at the time of this writing, constitute the Biloxi-Chitimacha Confederation of Muskogees (BCCM). After first being denied federal recognition in 1994, in a submission where Native residents from these communities were lumped under the designation of the United Houma Nation, BCCM formed in 1995 when these three Tribes came together to work toward obtaining federal recognition as tribal sovereign nations. The Pointe-au-Chien Indian Tribe adopted its name in 1995 for its federal recognition process. A BIA representative I spoke with, who was first assigned to the Tribes' case over twenty years ago, explained that the BIA *did* find that people from the three settlements were Indian, but now it was a matter of matching their specific tribal designation. Native residents explained how their elders told them they were Choctaw or Chitimacha, but for the most part they were raised as Indian, without any specific tribal affiliation.

In June 2004, the State of Louisiana acknowledged the Isle de Jean Charles, Grand Caillou-Dulac, Bayou Lafourche, and Pointe-au-Chien Indian Tribes, as well as the United Houma Nation, and granted them all state recognition.[11] However, state-recognized tribes are not recognized by the federal government as sovereign nations and do not have reservation land. Several tribal members repeatedly conveyed to me how state recognition provided very little and that they did not have the same rights and access to land, education, and health services as federally recognized tribes.

At the time of this writing, the Tribes have yet to receive federal recognition. This is in part because of historical injustices, such as being forced into isolation and lack of formal treaties or formal relationship with the federal government, making it exceedingly difficult to provide written evidence of a sustained political authority and descent from an historical tribe, which are some of the criteria required for federal recognition.[12]

Under the Indian Reorganization Act of 1934, the Office of Indian Affairs exerted the authority to create a process whereby the US government dictated tribal recognition. After formalizing the procedures in 1978, the government accepted tribes as federally recognized if they had engaged with the government through treaties, lawsuits, or policy enactments.[13] There are currently 573 federally recognized tribes in the United States. However, approximately one-quarter to one-half of Indigenous peoples in the United States and the US-occupied territories are not federally recognized.[14] Tribes whose lands were grabbed during the early colonial era and did not enter into any formal relationship with the US government through treaties were not automatically accepted.

Tribal leaders and members, scholars, and researchers have challenged distinct issues with the federal recognition process, such as criteria being applied inconsistently and the increasing level of proof required to meet individual criteria.[15] Not accepting oral traditions and requiring outside verification of tribal identity has created distinct disadvantages for many southern and eastern tribes in particular, who, faced with forced removal, employed survival strategies of avoidance and hiding identity.[16]

Federal recognition, which institutes a political and legal relationship between a tribe and the United States, affirms the sovereignty of a tribal nation and helps protect the tribe's political, legal, and cultural rights. However, it can also act to authenticate the US colonial authority and control over the particular tribal nation within a racialized discourse that primarily serves the interests of the United States.[17] The federal recognition process requires a tribe petitioning for recognition to submit documentation on its genealogy, culture, and history to the US Bureau of Indian Affairs (BIA) Office of Federal Acknowledgment, making this office "the unelected arbiter of Indian identity in many ways."[18] In the federal recognition discourse, tribal cultural identity is considered authentic if recognition is received,[19] implying that tribes that are not federally recognized are culturally inauthentic. Dominant ideologies that constrain, dominate, and deny them cultural citizenship are an act of violence and an assault on one's identity and sense of self.[20]

Obtaining federal recognition can become a source of pride and economic benefit but can also lead to increasing powerlessness. While trying to gain federal recognition and obtain resources for educational, economic, health, and elderly support programs, tribes can be negatively affected by the federal authority asserting control over a tribe's sovereignty and cultural integrity.[21] Several members of the three Tribes that are the focal point here voiced their frustration, discussing how the recognition process was designed to separate tribes and breakup tribal identity and cultures. Yet, opportunities for federal government grants or programs that assist tribes—including for climate adaptation and mitigation and environmental restoration—require federal recognition to qualify. Without recognition, tribes often do not have access to these resources.[22] And without recognition, tribes are not acknowledged by the US government as having distinct governing sovereignty as a Nation, which is more than a matter of legal and political rights.[23] Some tribal residents in coastal Louisiana felt that without recognition, a sense of their identity was being denied. For example, a Grand/Caillou Dulac Tribal member who had relocated about 100 miles northeast from Dulac showed me some headbands he had made with fake feathers to wear at powwows. Because he was not part of a federally recognized tribe, it was illegal for him to obtain eagle feathers, and he could not dance the eagle dance at powwows.[24]

To support the Tribes' work for their federal recognition process,[25] I tried to follow a paper trail from the library in Houma to research centers in New Orleans to the State Archives in Baton Rouge to the National Archives, the Library of Congress, and the Anthropological Archives in Washington, DC. At each stage of the research, a librarian or archivist would chuckle when I explained the type of information I was looking for, such as French and Indian trading records, noting that such information would be nearly impossible to find outside of the Five Civilized Tribes, referring to the Cherokee, Choctaw, Chickasaw, Creek, and Seminole Nations.

For the young adult tribal members who helped with the Tribes' research efforts, the struggle for federal recognition had been going on most of their lives. One distinct issue was that disasters continuously interrupted the Tribes' research. For example, Michele, a Pointe-au-Chien Tribal member who had relocated from Pointe-au-Chien about 15 miles north to Bourg, explained that several Pointe-au-Chien Tribal members were looking for how integration occurred in the local school system to show that "this school system, here in Terrebonne and Lafourche called us Indians, we're Indians they knew us as Indians. . . . [However,] we still haven't finished going through the school board minutes and since then we've had two storms."[26]

Another major issue the Tribes had in the recognition process was proving a cohesive community before 1830. The Tribes' ancestors migrated throughout the lower Mississippi Valley, and during the Indian Removal era they were either forcibly removed or forced into hiding, as illustrated in the next chapter on colonial legacies. Further, as several residents pointed out, their elders could not read or write, so how were they supposed to keep a written record of their history in a foreign language and outside their cultural norm of communication, particularly when they were fleeing from acts of genocide? The BIA was asking for written documentation that did not exist. The Tribes audio recorded oral histories and transcribed these records, but the BIA had thus far not counted the oral history tradition as evidential proof, placing the Western mode of research and documentation above traditional methods of passing along history and knowledge. And many elders had become weary of retelling their story over and over without results.

One possible avenue to support their submission is for the Tribes to prove land claims by tracing land holdings back to what the Spanish gave to the Tribes during the era of Spanish rule in the latter part of the 18th century. After Spain gained control of the region, Spanish colonial rulers informally agreed to protect the Chitimacha's territorial rights; however, extending these rights implied that such rights could also be taken away.[27] Further, these records, even if obtained, would be in Old Spanish and difficult and costly to have translated. Without financial resources to obtain and translate such records, it is nearly impossible to prove these land claims and other criteria necessary to obtain recognition. Furthermore, the Louisiana territory changed hands a number of times to different colonial rulers, as will be discussed in Chapter 4. The French colonists who took over after the Spanish did not recognize the Native groups' land claims, which continued to go unrecognized when the United States took over control of Louisiana.

Federal recognition also plays a role in the Tribes' ability to cope with disasters together as a unit. If the Tribes receive recognition, Chairman Chuckie of Pointe-au-Chien saw one benefit being access to resources to try to obtain property for Tribal members to have a safe haven to go to together during hurricanes and flood events. As he said, "Without federal recognition, hard to get funding. During or right after storm, organizations might contribute, but other than that seldom happens. With federal recognition would have bigger voice, especially during disaster times. Even with BP deal, would've had bigger voice and say so about what's happening if federally recognized," referring to the claims process following the 2010 BP Deepwater Horizon Oil Disaster, which will be discussed in Chapter 5.

In 2008, the Tribes re-applied for federal recognition but were once again denied.[28] Their resubmission process has since been on official hold due to continued hurricane recovery. At the time of this writing, the three Tribes are working on their resubmission.

Accumulating Disasters in a New Climate

Coastal Louisiana is losing land at one of the fastest rates of coastal land loss in the world. Containing approximately 41 percent of the nation's coastal wetlands,[29] coastal Louisiana is experiencing 90 percent of the total coastal wetland loss in the continental United States. Between 1932 and 2010, the region lost more than 1,800 square miles of land.[30] This equates to a decrease of about 25 percent of land area in coastal Louisiana since 1932, with 25 to 35 square miles of land disappearing every year.[31] More than 300 square miles of marshland were lost to Hurricanes Katrina, Rita, Gustav, and Ike alone from 2005 to 2008.[32] At the current rate of land loss, an area equivalent to Washington, DC, will be under water in less than two years.

The three primary land loss processes include erosion (removal of land by water action), submergence (increase of water level relative to ground surface elevation), and direct removal (removal of land by actions other than water).[33] There is a natural process of subsidence with the organic sediments deposited along the coast compacting, consolidating, and oxidizing. However, human actions—predominantly oil and gas industry activities—have caused nearly 70 percent of Louisiana's land loss.[34] In the Pointe-au-Chien area, the US Geological Survey compared aerial photographs and found that much of the wetland loss occurred between 1969 and 1974,[35] which follows the period of mass development of channels and canals for oil and gas pipelines (1950s and 1960s). Subsidence rates near oil and gas production fields are noticeably higher than geological rates of subsidence.[36] Local residents reiterated this issue, teaching me about how the land was sinking in part because of all the oil and gas resources being extracted from the earth.

In particular, throughout the past several decades, oil and gas companies dredged thousands of miles of marsh and wetlands along Louisiana's coast for pipelines and navigation, which caused considerable land loss, erosion, and extensive saltwater intrusion. The dredged canals become substantially wider as the current and hurricanes pull more marshland away. The construction of dikes and levees and damming of the Mississippi River by the US Army Corps of Engineers (USACE) and local levee districts, other flood control measures, and development for cypress logging and large-scale agriculture prevent sediment and silt from reaching the Delta, prohibiting the land from building up and replenishing. The number of hurricanes over the last fifty years, and the attendant flooding, is rapidly accumulating (Figure 3.12). As the land sinks and the tide rises, the roots of the trees are inundated with saltwater and no longer get enough oxygen. The dead trees signal that the land will soon be gone (Figure 3.13).

Sea level rise and intensified hurricanes compound the effects of erosion and subsidence to coastal Louisiana. The above processes are producing one of the world's highest rates of relative sea level rise—sediment subsidence combined with sea level rise—with an over 8-inch rise in the last fifty years, slightly faster than twice the global rate.[37] Recent data from the US National Oceanic and Atmospheric Administration (NOAA) predict the region will experience an additional 4.3 feet of sea level rise by the end of the twenty-first century.[38] The sea level rise

DATE	TROPICAL STORMS AND HURRICANES
1957 (June 27)	Hurricane Audrey
1964 (October 3–4)	Hurricane Hilda
1965 (September 9)	Hurricane Betsy
1974 (September 7–9)	Hurricane Carmen
1985 (October 29–31)	Hurricane Juan
1992 (August 26)	Hurricane Andrew
2002 (October 3)	Hurricane Lili
2003 (June 29–30)	Tropical Storm Bill
2005 (August 26–29)	Hurricane Katrina
2005 (September 24)	Hurricane Rita
2008 (September 1)	Hurricane Gustav
2008 (September 13)	Hurricane Ike
2011 (September 2–4)	Tropical Storm Lee
2012 (August 28–29)	Hurricane Isaac
2015 (October 20–24)	Hurricane Patricia

FIGURE 3.12 Tropical storms and hurricanes in and around the three tribal communities, 1957–2015.

FIGURE 3.13 Ghost forest and land loss near the three tribal communities. The dead trees on the left are a warning sign for the land loss soon to come, as shown, for example, on the right.

Source: Photo courtesy of Babs Bagwell, 2012.

and storms increase the salinity in estuaries and wetlands. What had been freshwater was now brackish, and what had been brackish water was now salty, impacting the fish that the Tribes and other subsistence coastal communities have relied on for sustenance and food security for generations.[39]

The loss of landmass and barrier islands to the south of the tribal communities has decreased the natural protection against hurricanes and storms. The communities have now become the shock zone, bearing the brunt of the wind and flooding, and other related storm impacts for communities farther inland. What were once small ponds are now lakes: some lakes have become large bodies of water merged into the Gulf of Mexico. Such extreme coastal land loss has resulted in the map of Louisiana needing to continuously be redrawn, with names of places and bodies of water vanishing from the map.

Climate Change amidst a Suite of Stressors

During a workshop in January 2012 held at the Knights of Columbus Hall (a Catholic-based fraternal organization) in Pointe-aux-Chenes, which is often used as communal meeting space, representatives from the three Tribes and Grand Bayou Indian Village, another tribe in coastal Louisiana, and local researchers, including myself, gathered together to discuss the effects of climate, weather, and other environmental changes on the communities.[40] Tribal members discussed how they experience the impacts of a warmer climate: many agreed that it stayed warmer longer and winters were shorter than before. The hotter temperatures prevented the plants and trees from entering their customary dormant season that was needed for good production and plant health. Their social interactions and sense of community were diminished as people stayed inside more with air conditioning on and kept windows shut. The hotter temperatures also harmed their livelihoods. Shrimpers now needed a cooling system for their catch or had to come in sooner because of refrigeration concerns in the heat. Based on the workshop's discussions, I created a report with the tribal leaders to submit to the US National Climate Assessment.[41]

Going through the workshop and report process provoked my thinking more carefully about the way researchers, scientists, and journalists interact with local people about climate change and the way people's words are then communicated and presented back to the public. For example, during an Isle de Jean Charles story circle the week before the workshop, a couple tribal members discussed hotter temperatures:

Pierre: It was colder then than today.
Louis: Oh yeah, used to have snow and freezing.
Pierre: Three or four days iced up.
Louis: Even the weather changed for down here!

If I only listened for what I wanted to hear, that the hotter temperatures are directly because of climate change, I would have missed what they talked about just a moment later when I asked what had been the cause of the changes:

Pierre: We getting close to the Gulf, that's why. When you get closer to the water, the cold don't go as far . . .

Louis: Once they started digging all them canals and all that, that's when the salt-water started coming in. That's when we started losing all of the ground, all of our trees and everything. That was all through the oil companies.

Considering the complexity of environmental change, explicit attention needed to be paid to the reasons local residents gave for experiences such as warmer temperatures, like the loss of land from digging canals, which caused the warm Gulf water to creep in. Discussing her community experiencing more flooding, Marlene (Grand Caillou/Dulac) noted, "Everybody keeps saying it's global warming and all that stuff. But you have to remember they have oil and gas under this earth and they're pumping all of that out. You put something in a glass with a full glass of water, and then you start pulling out the water, that top part is going down. That's what's happening now."[42]

Following several centuries of displacement and oppression, some tribal members feared that climate change could be used as an excuse to move people to supposedly get them out of harm's way but in fact with the goal of developing the coast for tourists, large-scale commercial fishers, and the oil industry. Several residents discussed rumors about outside developers moving in to buy up the land, forcing local people to move.

Their concerns were not without precedent. Some governments, in the name of moving people out of harm's way, have permanently moved some communities who were temporarily displaced due to disaster events, at times conveniently making way for private development. For example, after the 2004 Indian Ocean tsunami, the Maldives' government announced that residents needed to move to one of five islands designated as safe zones, thus clearing entire areas for tourism and resort development.[43]

However, some local residents in coastal Louisiana had in fact started to recognize climate change among the suite of stressors they were experiencing, noting the hotter temperatures and increasing intensity of storms, but with so many co-occurring events and habitual disasters, it was often difficult to isolate any one cause.

Habitual Disasters and Relocation

From 2005 to 2015, the communities severely flooded nearly a dozen times from eight hurricanes and storms, most recently from Hurricane Patricia in 2015, in addition to king tides and south winds.[44] The accruing, habitual disasters have forced more and more people to relocate because of severe flood damage to their houses, lack of resources to elevate their houses, people being tired of rebuilding, or the skyrocketing flood insurance rates, which could cost over $25,000 a year for at-risk homes in the parishes where the Tribes are located.[45]

Some left to pursue other job opportunities because of the loss of fishing-based livelihoods, a result of a combination of severe environmental changes and a changing seafood industry flooded by lower-priced, subsidized, pond-raised imports, the effects of which have been unfolding for over half a century.[46] Others left because they had difficulty getting to work with the roads flooding during high tide. Some people left to seek better educational and employment opportunities or because they married someone outside the community. Wanting to stay together and maintain their community and culture, a number of people coped with relocation by remaining as close as possible to a familiar landscape and to other people from their Tribe, clustering together on slightly higher ground farther north.

Those who had relocated often expressed a strong desire to return. Marlene (Grand Caillou/Dulac) expressed this sentiment as we drove slowly up and down the bayou running alongside Dulac.[47] We stopped where she grew up and got out of the car. I took a picture of her looking into the camera with her hand pointed back to display the now empty lot along the bayou where her house once stood, a dead tree with its ghostlike limbs rising up behind her and the area where her dad once grew gardens now covered with grass above ground and salty water below. I asked her if she would ever move back here. She said absolutely if it could be saved.

Livelihood, Health, and Socio-cultural Effects

For people who have dwelled in a place for generations, even having to move short distances can have significant effects. As Chief Shirell (Grand Caillou/Dulac) shared,

> With us having to move away and break apart, we lose the traditions. We lose the culture because we can't stay within our community and practice our beliefs. . . . You know, when I lived in Grand Caillou, my area was American Indian. That was my family. I was brought up very heavily in my traditions and our culture and who I was and to be proud of it. And you know, we prayed. And when we were sick we were brought to our great-grandmother. And that's what it was. And we went to our grandmother and we sat down and we heard her radio playing and dad drank coffee. And you know, um, that was what we did. And when we have to move, we lose it. And those things continue and continue and continue as long as you're able to stay in your area because that's what you're used to doing, it's practice. It's like breathing or tying your shoe. . . . We don't have that anymore, you know. It hurts.[48]

Serena, a Pointe-au-Chien Tribal member who relocated to Bourg, about 15 miles north of Pointe-au-Chien, described to me the difference between living in the two places even though they were so geographically close. She provided an explanation that an outsider like myself needed to understand the significance of relocating even a few miles:

> What makes it unique is the fact that over all these years, the community has been in this bubble down there. It's different. From being here, just to go ten miles down there, everybody is really close knit. If you're going down the road, they're going to wave. There's going to be a group of people standing on a dock somewhere, all talking and just being together. And you can just walk up and they'll treat you like you're one of them. And there's not many communities still like that. The people down there are just so welcome.[49]

Comparing being in Bourg to Pointe-au-Chien, Serena said, "Just from being here to there, it's that different."

What Chief Shirell and Serena both described relates to what disaster survivors often report—loss of livelihood, health, access to education and services, resources and resource exchange practices and reciprocity, as well as diminished cultural practices, and fragmented and disrupted social networks.[50] Further, one's social network aids response and recovery efforts and long-term adaptation through both tangible (e.g., shelter) and non-tangible (e.g., emotional) resources. Such support is often through informal exchanges at the community level that are hidden from direct outside view.[51] This relates to how people cope with crises in their everyday lives, such as the reciprocity of informal exchange through childcare or money lending. The effects accumulate as disasters often send shock waves through social systems that can inhibit responses to the very challenges the disaster presents and intersect with increasing challenges experienced in people's everyday lives. As disasters become repeat processes, and people are left to still be recovering from one disaster when another hits, the effects become more and more heightened, as the systems they once relied on for recovery and long-term adaptation are already diminished.

The tribal communities' social networks are particularly deeply established after spending decades socially, economically, and politically isolated. The isolation—and subsequent incursion and land grabs—helped to foster close-knit ties and also wariness about any outsiders coming into their space. Utilizing their networks, information quickly spreads throughout each community. For example, I would often come across a local resident I hadn't seen for a couple weeks, but the person would note in passing what I did and whom I was with the previous day. With an increasing number of unknown entities and actors coming in and out of the communities, residents kept a close eye on who was there and for what purpose. There was an increasing sense of threat from two types of flooding—that of the encroaching water, and that of more people flooding into the area and competing for resources and control of access to resources.

Retaining fishing livelihoods and community rituals is increasingly difficult for the three Tribes: people no longer gather and share as they once did, and the everyday traditions and cultural practices that tied families and communities together have gradually diminished. It was once common to see families gathered together on their fishing docks, but with so many people forced away from a fishing-based livelihood, the time together on the boat or preparing seafood and sharing livelihoods and stories has been severely reduced.

The current dynamics contrast to how the communities previously functioned. Expressed during a story circle at the house of two elders—Pierre and Marie—on Isle de Jean Charles, tribal members discussed what it was like growing up on the Island:

Maurice: When they had a Depression, the people down here didn't know they had a Depression because nobody here had any money. They just trade fruits and vegetables and fish.

Pierre: They'd get whatever they wanted.

Louis: They lived on the land and like I said, when they was in the Depression we was not. Everybody helped each other.[52]

A number of local residents expressed how when they were growing up, they ate from the land and waters and shared with each other. However, I witnessed how residents now had to seek food and materials from elsewhere. Saltwater intrusion and contamination from heavy metals made personal and community gardens nearly obsolete, and, particularly since the 2010 BP Deepwater Horizon Oil Disaster, fishers were bringing in less catch. Local residents now relied more and more on processed foods they had to buy from the grocery store. This created an additional economic burden, as they had to purchase resources necessary for physical survival that were previously harvested or caught for free. With these changes, the sense of sharing had diminished. Growing up, they used to build boats and houses together and trade crops and seafood, but, with more people relocated and the soil and water contaminated, they were losing a core cultural value, as well as economic and physical, mental, emotional, and spiritual health benefits.

Besides the stress of having to pay for what they once obtained freely, several people shared how they had gained weight from their change in diet, especially since the BP Disaster. They were substituting their seafood catch with meats they had to buy at the store. For example, Robert, a local fisher, who had relocated from Pointe-au-Chien about 10 miles north to upper Pointe-aux-Chenes, noted how he had gained a lot of weight since the BP spill; instead of seafood, he was eating more meat and food that he was not used to eating. He used to be able to walk outside and catch whatever he wanted to eat. But now, as he articulated, his catch was diminished and he had no livelihood.

Severe erosion and intense saltwater intrusion had killed the majority of the communities' trees, traditional and medicinal plants, and gardens, and the trapping grounds, along with the animals, were gone. For example, 25 percent of medicinal and edible plants had disappeared from Pointe-au-Chien, and 40 percent were increasingly rare and becoming more scarce.[53] There was only one substantial garden left in the community, maintained by an elder and her sons. However, this garden was also at risk, flooded under two feet of water after Tropical Storm Lee hit in 2011.

The Tribes used to have *traiteurs*, tribal members gifted with the knowledge of prayer and herbs, who would go into the woods next to their houses to find medicinal herbs for treating illnesses. Now, Pierre, an elder from Isle de Jean Charles, told me, "We don't have no medicine no more. The saltwater killed all the medicine."[54] Residents have to pay to go see Western medical doctors miles from their communities, and much of the knowledge about medicinal plants has been lost.

The erosion of traditional medicinal and plant knowledge is due not solely to environmental changes but also to the knowledge being devalued. The younger generation has been assimilated and integrated into a US society that values only the solutions of institutionalized medicine, which "conceptualizes disease as a discrete entity,"[55] as opposed to the Indigenous art and tradition of treating and healing.

With the loss of cultural resources like palmetto leaves for basket weaving, the relocation of much of the younger generation, and the introduction of new technologies, traditions like the art of carving pirogues, making cast nets, and basket weaving are being lost. Mary, an elder and basket weaver who grew up in Isle de Jean Charles but relocated years ago to upper Pointe-aux-Chenes, would like to see the tradition passed down, but there were only a few people who still knew how to do it. Therefore, "when we go, it goes too."[56]

"When I Go Back Now I Go Kind of Blank"

Talking on the phone one day, Victor told me how much he would love to come back to Isle de Jean Charles from where he lived about 140 miles northeast in Mississippi. He described the pictures he still had of all the trees that were once near the house where he grew up and the cattle for which his family cared. But after the saltwater came in and the land eroded, the trees died and it was no longer possible to sustain cattle on the Island. He talked about how quiet the Island used to be, with the only sounds being buyers coming to pick up crabs and shrimpers going to sell their shrimp at the factory in Pointe-au-Chien and another shrimp factory farther down the bayou. Now, large trucks passed regularly as outside recreation users went to the south end of Isle de Jean Charles to launch their boats from the private marina or to go to their fishing camps. He had a lot of good memories from Isle de Jean Charles, but, as he expressed, "When I go back now I go kind of blank."

From trapping on the same lands and fishing in the same waters as their grandparents and generations before them, the accumulated knowledge of generations pursuing the same subsistence activities across the same water- and landscape tied people together. Subsistence activities helped shape people's identities. As anthropologist Thomas Thornton (2008: 119) articulated in his ethnography about the Tlingit people of Alaska, "to procure food—to subsist—is quintessentially to dwell, to gain sustenance and 'real being' from places." For tribal communities in coastal Louisiana, time was not experienced by the 12-month calendar as much as by seasons and cycles for gardening, trapping, hunting, and fishing. Time for these activities, however, is now designated by the Louisiana Department of Wildlife and Fisheries, predominantly due to large-scale commercial fishing depleting resources. People have thus become alienated from the practices that united them across generations.

People's place attachment stems from both their own experiences and multigenerational knowledge and wisdom of the physical environment in which they carry out their livelihoods and cultural practices and in which their memories, narratives, and stories told over generations come to life. Anthropologist Setha Low (1992) explained the relationship between people and the place they occupy as "formed by giving culturally shared emotional meanings to a particular space or piece of land that provides the basis for the individual's or group's understanding of and relation to the environment." However, with the rapid land loss and accruing environmental changes, tribal members were unable to pass down some of their traditions to their children and sense of belonging to that particular place, resulting in distinct physical, cultural, and spiritual loss. As Chief Shirell explained,

> I've always wanted to be able to take my kids and play with them in the woods like I did so they could see what it was like to be an Indian kid growing up in that setting because you become one with Mother Earth, you respect her. You get to see all of her beautiful gifts. It's important. And with everything that's happening I can't give them that. I think that's what hurts me the most. As it was given to me by my parents, just like it was given to them by theirs, and I can't give it to mine. And I can't give it to mine. And they won't be able to give it to their children. It's a hard reality to face.[57]

Similarly, Gabrielle, who was born on Isle de Jean Charles but whose family relocated about 20 miles north to Houma when she was a child due to severe and perpetual flooding on the Island, shared,

> If we lose the Island, we lose what brings us back to it. And that's the idea that that was our place. It was our place. Everybody else can say, the government considered it uninhabitable, and we took it and inhabited and we made it our place, and now it's gone. It's going. And if it goes we'll no longer have our special place. That's the one thing that keeps us together as a community, as a reservation, is we had our place. We don't have our place anymore. We have no place.[58]

Gabrielle's insight suggests the kind of spiritual blow that people endure as they watch their homelands disappear. This blow is just as fierce for many who have already relocated and fear what it means if they lose their homelands entirely. People's everyday life experiences, memories, and events are carried out in a specific place; thus, the attachment developed to that place creates a sense of their own identity.[59]

To some, both the landscape and the community had become unfamiliar, and they expressed distress over returning. Crystlyn, who lived with her family in Grand Caillou, noted this distress when we drove south from Grand Caillou to Dulac so she could show me the street where she grew up.[60] After habitual flooding, many Grand Caillou/Dulac Tribal members had moved farther north, and new people, including drug dealers, had moved in, changing the dynamic of the community and place. As we drove along, she pointed to the remnants of shrimp factories and other businesses and stores. She showed me where her grandmother's house once stood on Shrimper's Row. She said it used to be the most beautiful place but now she did not know the people there anymore.

Marlene told me when we drove around Grand Caillou and Dulac how there used to be so many stores in Dulac, but now there was hardly anything. Stores started closing after Hurricane Andrew hit in 1992 and people started relocating. More stores closed following Hurricanes Katrina and Rita in 2005, with more people relocating and storeowners no longer able to afford the high flood insurance rates to maintain the stores.

During a story circle at Geraldine's house on Shrimper's Row in Dulac, Marlene and Geraldine discussed the closing of local businesses in the community:

Marlene: It was because of hurricanes. And I mean, people get the shrimping business was really bad. . . . If they had a lot of shrimp, they only got paid very little for their product.

Geraldine: Then the oil field went down.

Marlene: Yeah, so people just didn't have the money and you support your local community, you know local businesses, but when you have no money you can't support your business, so they all went belly up.[61]

The women pointed to co-occurring adverse events—hurricanes and the collapse of the local oil and seafood industries—that led to so many businesses closing in their community. The loss of stores had more than an economic impact. Crystlyn described to me what it meant for the local businesses to close: "Well, we had a grocery store. Freakin' best meat you could buy! Oh man, I miss that place so much. And like you'd go in and see your cousins are working there and you knew everybody in the store."[62] Local residents had spent generations isolated down the bayous and were accustomed to knowing everyone and everything around them; this was an important aspect of community life. Feelings of dislocation surfaced as the sense of knowing their neighbors and landscape shifted, as their social memory of place and current reality were disjointed, and what was once known became unfamiliar.

"It's the People That's Made the Island": Social Memory of Place

I sat with Gabrielle in her house in Houma while her children played in the other room. She talked about the important role of storytelling in maintaining the Isle de Jean Charles Tribe's culture: "The stories, those are the things we're going to have, so we have to keep those alive in order to keep the memory of it alive and to keep the culture alive because if not, when the Island goes the culture's going to go with it." She discussed what it would mean if the Island completely eroded away:

What will the people have? It's erased. That's it, they won't have that. They can't say, oh okay we're going to go back to mama's house and cultivate the land and do this and that. You're going to miss that whole sense of who we were, the people playing in the road, teaching your kids to swim before they could almost walk, you lose the art of fishing, you lose those things.[63]

Tied to distinct cultural components, in the above passage Gabrielle points to the social memory of place. Historian Geoffrey Cubitt (2007: 14–15) explained social memory as "the process (or processes) through which a knowledge or awareness of past events or conditions is developed and sustained within human societies, and through which, therefore, individuals within those societies are given the sense of a past that extends beyond what they themselves personally remember." The past not only produces the present, but the present also produces the past, in which memories and re-telling of the past are selected and made significant through what makes sense for the present.[64]

An example of this occurred during a Grand Caillou/Dulac story circle at Geraldine's house on Shrimper's Row in Dulac, when Geraldine started telling a story about one of her favorite memories: "I remember [dad] coming home with sacks of nutrias and muskrat on his back, carrying them from the swamp. Somewhere, we would get there and skin these animals and dry so he could sell them up the bayou." Geraldine and Marlene then started talking about oil companies destroying the land, which transitioned Geraldine back to the memories of her family working together trapping and skinning:

Geraldine:	And that's gone. That's been gone.
Marlene:	Yes, even the muskrats. There's no more of that.
Geraldine:	And that's how we lived.
Shirell:	Simple, simple life. You know, an honest, hardworking simple way of life and nobody bothered anyone.
Marlene:	And that's been in the making I think. The United States deals so much with the countries, you know the other countries.
Shirell:	Foreign trade.
Marlene:	Yeah, foreign trade that they make deals with them to destroy the fishing industry down here because that's been talked about for years.
Shirell:	If I read it was an import, I'm not eating it, I'm sorry. Have you read what's in an import?
Marlene:	It's sad, but that's what's happening.[65]

Starting from trapping and livelihoods, the women's conversation traveled through the effects of environmental changes and loss of subsistence activities to the consequences of globalization on their community's livelihoods and way of life, pointing to how the local shrimping industry had drastically declined in Louisiana over the last several years following the significant increase in imported shrimp.

Tribal members often connected the past and present together by linking experiences and knowledge across generations. Social memories created a link between people and between people and the landscape by forming a common, shared community narrative in which people shaped the landscape with which they identified. For example, when I stood with Antoine, a Pointe-au-Chien elder, on his back porch in Pointe-au-Chien looking out at the small clumps of remaining marsh between his house and the couple miles across to Isle de Jean Charles, he explained how "we used to be able to walk to the Island from here." He said "we," but when I asked if he was ever able to do that, he said no, that was more what his grandfather did.[66]

A particularly revealing moment came during a conversation with Henri and Josette, both from Isle de Jean Charles, in their home in Grand Bois about 20 miles north of the Island. I brought up the topic of when they left the Island in the 1970s.

Henri looked down at the floor and then, peering over the top of his eyeglasses, asked me about my family's movement. I told him where my extended family lived and about my own personal physical movement. When I finished talking, he sat up a bit more, signaling that he was ready to respond. He breezed past his own personal life but described in detail the history of his people and how they came to live down the bayous. Josette, who was now in the kitchen making shrimp patties for lunch, shouted over her shoulder, "It's the people that's made the Island, not the Island's made the people."[67]

At times I had difficulty keeping up with whether people were referring to the past, present, or future. Tribal members weaved stories together from the Trail of Tears, to their own ancestors being displaced and escaping down the bayous, to the racial segregation they lived through themselves, to the last half-century of environmental destruction and land grabs, to what would happen if measures were not taken to stop the flooding and restore the land. Social memories of the past were constructed through what tribal members were told by their ancestors about being Native, escaping down the bayous, and adapting to their new environment.

People's memories are part of their community's narrative, which is shaped by their connection to place and the inclusion, belonging, and connectedness to the past in that place, as well as the past that directed them to live in that place.[68] The communities' social memory was not just of their current geographic location but also the story of how their ancestors came to be there. The loss of place—placelessness—can signal a severance with past generations and a loss of cultural identity and practice for present and future generations; it is not a material element that can be quantified and compensated. Thus, the vast and rapid land loss created a sense of dislocation, for both residents who had relocated and those still in place.

Solastalgia

When I asked local residents about land loss, they often started by telling me the cause of such loss. Then, without my provoking, they typically followed with a personal story of somewhere they used to go, a place that was special to them, where they would trap or garden, spend time with family members, keep animals, or pass by on their boats, and about how the place they were referring to was now gone or disappearing. For people who spent their lives navigating coastal Louisiana's intricate web of waterways, as the water- and landscape changed, the land loss was not just physical but affected their sense of place and belonging.

People can experience a sense of dislocation as places they identify with and where their memories were lived out become unfamiliar. This sense of dislocation evokes feelings of what philosopher Glenn Albrecht and others (2007) coined as "solastalgia," which is "the distress that is produced by environmental change impacting on people while they are directly connected to their home environment." In short, solastalgia is "a type of homesickness one gets when one is still 'at home.'"[69]

Several tribal residents expressed this sense of dislocation. For example, one day driving along Bayou Pointe-au-Chien, I passed by Theodore, who lived on Isle de

Jean Charles, standing alongside the bayou next to his oyster boat. He invited me onto his boat. He pointed in both directions and said how much land there once was going both ways. He talked about where he used to oyster nearby. The canal used to be about 300–400 feet wide, but now it seemed like it was about a mile wide. He pulled out his satellite radar device. His kind, playful eyes turned to me, the deep-set wrinkles crinkled in his leathered face. He showed me on the device the surrounding places that used to be land and were now water, ponds that had become lakes. He described how he used to be able to navigate these waters with his eyes closed but now even the navigation technology he had couldn't keep pace with the disappearing land.[70] A few days later, Jean, an Isle de Jean Charles elder, expressed a similar sentiment describing where he used to travel by pirogue (small, dugout boats), noting, "Now it's hard to distinguish where those places are."[71]

Similarly, as I passed through a wide canal south of Pointe-au-Chien on a boat with Theresa, she talked about the dense forest that used to cover both sides of the bayou. Looking out at the dead trees that now overwhelmed the landscape, she pointed to where her grandparents had lived, which was now rapidly eroding, and described taking the boat down the bayou when she was growing up. The passage was so narrow that she could pull the grass on either side of the boat. Now, it is a wide canal, and the trees that were once alive and full are now skeletons.[72]

Many residents described their community as scattered and felt a sense of dislocation with so many community members having relocated. As Regina, from Isle de Jean Charles, described:

> They've got a space that's got no houses at all, a big space, where it used to be inhabited. Lots of people. And now, oh my God, it's like a ghost town. It was like a war zone when the hurricane came. Now you look at it and you hardly see any cars pass, you hardly see any children here. A few people here, a few children. Like when [Charles and I] first met, there were kids everywhere, kids coming out of the woodwork, but that's what made the community.[73]

She went on to discuss the meaning of home, "the place you were born. That's your station so to speak. But it's more than just a place where you were born. Your ancestors were here. The people before you were here. The genealogy, the tradition, everything, this is where you belong right here." But with so few households left on the Island, Chris, a lifelong Island resident, defined the way the Island now looked: "It's like the skeleton of the body."[74]

Tribal members often expressed feeling displaced through their feelings of mourning for a lost home. Chief Shirell described the concept of home as "it's what our culture and heritage is all about, it's where we live, it's our family, our friends. It's everything about us. Home's not a house, home is your community. It's where you grew up, it's where you want to grow old and die. It's just where you want to be."[75] The concept of "home" refers to a place where people are tied together through relationships formed by shared interests or beliefs, where their identities had been formed through knowledge and understanding of the local landscape, history, culture, politics, and economics.[76] As the shared home where people's memories are embedded washed away, a sense of dislocation, alienation, and uncertainty persisted about what this meant for the future of each tribe, their culture, and whether anything would be done in time to restore the land and community. One of the ways these feelings were most highlighted was through people's narratives about the loss of trees.

"If You Could Talk to That Tree It'd Probably Tell You a Few Stories"

When describing the changes in the landscape, one of the most prominent issues tribal members raised was the loss of trees, which had been such an integral part of the landscape and people's memories about their community. As Chief Albert said when we drove around the Island, "If you could talk to that tree it'd probably tell you a few stories."[77] The loss of trees represented the changes in not only the physical environment but also the community. People often described the dead trees as ghost forests and the communities as skeletons.

The dead trees dotting the landscape were reminders of what the landscape once looked like, what the community had been like, and what processes unfolded during the past few decades. For example, I stood with Victor at the house on Isle de Jean Charles where he had once lived before moving away due to flooding. Standing on the deck of his raised house looking across the open water at the oil wells nearby, he said, "We get a little check every month for that, but it's not worth, look what they've done to the, there was oak trees all over there. It's all gone."[78]

Continuing our conversation, Victor and I drove the approximately half-mile to the south end of the Island where he grew up. He wanted to show me the tree that was there the last time he went down there. As we drove I asked him how long it had been since he had been to the end of the Island. He said, pointing to landmarks outside the window,

> I don't really care to go there anymore. But I want to bring my grandkids over there and take a picture of that tree. . . . We used to live where that old cement slab is over there. That was the old cattle fence. That's the tree, oh shit! There's nothing left! That's the tree I'm talking about. My house was over there. . . . Look, they still got a little piece of oak tree over there, it's gone, but yeah, they blocked this up. You'd have to wear boots to go in the back. That's the canal I was talking about. They put the oil well right there where the oak trees were at, on other side of the bayou.

Back on the elevated porch outside the house, we watched the sun set. Victor continued to look out across the water. I asked him how it felt coming back to the Island for a visit. He said,

> It's not the same. Like in the back over there, look at that, all you see is lake, you just had that bayou at one time and now all you see is lake in the back.

See that line of trees there? You'd see trees all over. And the oil company kept on cutting and cutting. And the more you cut the more you sink. See like in Pointe-au-Chien over there, when they made that levee on the crossroad over there, daddy told . . . the police juror, he said you're all digging your own hole. When they made that they wanted to flood us down here. And instead of going out outside over there and making the levee, they would've protected everybody.

Victor transitioned from the loss of trees, to the oil industry cutting through the marsh to lay pipelines, to the perspective that government officials made flood protection decisions that specifically discounted the Island. The loss of trees in his narrative came to symbolize the lack of flood protection and restoration activities in and around the community, as well as the larger economic and political structures at play.

The Disasters Continue to Unfold

In late August 2012, I sat with Pierre and Marie, two elders from Isle de Jean Charles, at their kitchen table and watched the Weather Station's tracking of Hurricane Isaac on the television. Pierre repeated some numbers out loud that he saw flash across the screen. He went in the other room and came back with a map of the Gulf, Caribbean, and Central American region that someone hand-drew and gave him in 1987. He had used it to track the storms ever since. He looked at the latitude and longitude numbers shown on the television and marked an "x" on the map, the latest in a line of "x's" he had marked since the day before, tracking the hurricane's route. I asked them if they would evacuate or stay. They used to be protected from storms by land and trees and didn't have to evacuate, but this had changed in recent years. Marie did not want to leave, but Maurice, their son, was insisting they had to evacuate and stay with his family in Houma.

Across the street at Renée's, her grandchildren and great-grandchildren played under the house. Her daughter asked me how I enjoyed crabbing last week with a Pointe-au-Chien resident, signaling they knew what goes on in the area. I went up the stairs of the elevated house to see Renée. The kids wandered in, occasionally filling a bowl with spaghetti and *roux*, the basis for much of Louisiana cuisine. Renée said she did not want to leave but would probably end up evacuating to her son's house in Gray, about 35 miles northwest of the Island.

I drove back across the Island Road and stopped at the traditional palmetto hut Pointe-au-Chien Tribal members had recently built with their youth to teach them about how their ancestors had lived.[79] Pointe-au-Chien's Chairman Chuckie Verdin and his parents were there loading up scrap wood into his truck. He wanted to put webbing on the hut to protect it from the hurricane winds and waters, but there was not enough time; he had already been called to go back to work the next day on the tugboat.

I arrived back at our camp to a feast of boiled crabs and shrimp a neighbor left for us. That evening, my husband and I walked down to the south end of

Pointe-au-Chien. I watched local shrimpers pull up their land net, full of small crabs and sardines, but hardly any shrimp. The catch was still greatly diminished since the BP Deepwater Horizon Oil Disaster two and a half years prior.

As the red glow of the sun peaked through our window the next morning, I wandered over the bridge across the bayou to go down Oak Pointe Road, which designated the Pointe-au-Chien Indian Tribe's territory. The morning air was still amidst the quiet bustle of people loading boats up on trucks to transport them to higher areas up the bayou. Small boats traveled down the bayou to collect the wiry red, green, and yellow crab traps, which soon started piling up along the sides of the bayou. Oyster sacks were loaded onto trucks. Docks were emptied.

I drove the couple miles to the Island and saw Rebecca (a lifelong Isle de Jean Charles resident) and her daughter sitting in front of their trailer. As I got out of the car, Rebecca started to cry. After losing her house to Hurricane Gustav in 2008, she worried she would now lose the camper that she lived in. A few houses down, car trunks were open, filled with duffel bags and backpacks. Farther down the Island, Renée's children and grandchildren were packing up underneath the elevated house. They had indeed decided to leave.

I picked up Chris and his great-niece and great-nephew—Rebecca's grandchildren—and along with their two dogs we piled in the car. Chris said he decided to leave because of the possible tornado activity, especially of concern without any trees or land to the south to protect them anymore. But he planned to come back right away. He joked a lot but grew more quiet and serious as we drove into Houma and I dropped them off at the house of one of his sisters.

Back at our camp, my partner and I packed up the car and drove 30 miles north-west to a friend's house. As we drove away, I yelled across Bayou Pointe-au-Chien to a friend. She was on the fishing dock in her Cajun reeboks (the local slang for white rubber boots) preparing some crabs. She shouted back, "Hurricane party!"

Nearly four months later when I returned for a visit, the impacts from Hurricane Isaac continued to unfold (Figures 3.14). Rebecca and her two grandchildren were still living at Chris' house while their camper, which had a big hole in the middle and water flooded in, was repaired. Visiting with a tribal community a couple hours' drive to the east, residents had been able to return only days before. Piles of trash were lined up behind the houses from all the mud and debris. Some shared their survival stories, and about how they had missed a full shrimping season because their boats were busted from the storm.

Such cycles—disaster, recovery, prepare, disaster, recovery, prepare—are nothing new for the Tribes. With the history and continuation of so many accruing disasters, such events could be re-labeled as part of a "legacy of atrocities."[80] The following chapter illustrates that what the Tribes are experiencing today is a continuation of past policies: from colonial policies that forced the Tribes' ancestors to relocate, to land grabbing by oil and gas corporations and land developers, to current government restoration and flood protection plans that discount the Tribes' lands and resources.

FIGURE 3.14 Flooding from Hurricane Isaac, Isle de Jean Charles.

Source: Photos by author, 2012.

FIGURE 3.14 (Continued)

Acknowledgment

A special note of gratitude to Marlene Foret for contributing her story "Then and Now: A Tale of Life on the Bayou" to this book and for sharing her wisdom and story with me.

Notes

1 "Then and Now: A Tale of Life on the Bayou," digital story by Marlene Foret, Grand Caillou/Dulac, audio recorded on August 3, 2012.
2 Conversation with Chief Shirell Parfait-Dardar in Dulac, Louisiana, May 3, 2012.
3 Solet (2006).
4 Digital story, audio recorded on August 3, 2012.
5 A series of general education development tests, equivalent to a high school diploma.
6 Gremillion (2004: 67); Saunt (2004: 128).
7 Jackson and Fogelson (2004); Morris (2012).
8 CPRA (2012).
9 BIA (2008b).
10 US Census Bureau (2000, 2010b).
11 Dupre (2004).
12 Miller (2004); BIA (2008a). For recent revisions on the regulations governing the recognition process and criteria, see DOI (2015).
13 Roth (2008); Klopotek (2011).
14 Barker (2011: 28).
15 Klopotek (2011).
16 Miller (2004: 58).
17 Barker (2011: 6); Klopotek (2011: 2–3).

18 Klopotek (2011: 1).
19 Barker (2011: 28).
20 Ong (1996).
21 Klopotek (2011: 23).
22 Redsteer et al. (forthcoming).
23 Roth (2008); Rising Voices (2013).
24 Federally recognized tribal members can use real eagle feathers for cultural or religious purposes.
25 I was asked to help with research by Isle de Jean Charles and Grand Caillou/Dulac Tribal leaders.
26 Conversation with Michele in Pointe-aux-Chenes, Louisiana, December 20, 2012.
27 Hoover (1975).
28 BIA (2008a, 2008b).
29 Turner (1997).
30 CPRA (2017).
31 Couvillion et al. (2011); NOAA (2013).
32 Couvillion et al. (2011).
33 Penland et al. (2000).
34 Ibid.
35 Morton et al. (2005).
36 Morton et al. (2006).
37 Karl et al. (2009); NOAA (2012); Melillo et al. (2014).
38 Marshall (2013); Osborn (2013).
39 Maldonado and Peterson (2018).
40 The workshop was part of a week-long set of activities, Developing a Coastal Louisiana Tribal Conservation Advisory Council: Wisconsin Tribal Conservation Advisory Council's (WTCAC) Outreach Assistance to Louisiana Tribes, Pointe-au-Chien, January 23–27, 2012.
41 Coastal Louisiana Tribal Communities (2012).
42 Conversation with Marlene Foret, driving around Dulac, Louisiana, April 12, 2012.
43 Klein (2007: 505, 507).
44 Maldonado and Peterson (2018).
45 Wilson (2013).
46 Starting in the 1950s, the demand for shrimp surpassed the production levels of domestic shrimpers, and shrimp imports began to rise in the US market (Harrison, 2012). With the continued growth in demand for shrimp, foreign producers explored other methods to harvest shrimp. Aquaculture-based shrimp farming, funded by some of the same US-based mega-agribusinesses that pushed for large-scale, intensive agriculture, such as ConAgra, emerged as the new dominant form of shrimp production (Harrison, 2012). As these processes emerged based on a neoliberal model of export-based production, local residents where the shrimp farms were located, predominantly in Asia and Latin America, were often displaced and faced loss of livelihoods and land rights (Stonich and Vandergeest, 2001).
47 Conversation with Marlene Foret, driving around Dulac, Louisiana, April 12, 2012.
48 Conversation with Chief Shirell Parfait-Dardar in Chauvin, Louisiana, December 20, 2011.
49 Conversation with Serena in Bourg, Louisiana, August 2, 2012.
50 Peterson and Maldonado (2016).
51 Maldonado (2016b).
52 Story circle, Isle de Jean Charles, Louisiana, March 15, 2012.
53 Kachko (2013); Kachko et al. (2015).
54 Conversation with Pierre in Isle de Jean Charles, Louisiana, May 25, 2012.
55 Singer and Clair (2003: 424).
56 Mary, Pointe-aux-Chenes, December 21, 2012.
57 Conversation with Chief Shirell Parfait-Dardar in Chauvin, Louisiana, May 16, 2012.

58 Conversation with Gabrielle in Houma, Louisiana, July 29, 2012.
59 Burley (2010: 41).
60 Conversation with Crystlyn Rodrigue, driving around Dulac, Louisiana, June 11, 2012.
61 Story circle, Dulac, Louisiana, January 8, 2012.
62 Conversation with Crystlyn Rodrigue in Houma, Louisiana, March 14, 2012.
63 Conversation with Gabrielle in Houma, Louisiana, July 29, 2012.
64 Cubitt (2007: 27–28).
65 Story circle, Dulac, Louisiana, January 8, 2012.
66 Conversation with Antoine in Pointe-au-Chien, Louisiana, December 20, 2012.
67 Conversation with Henri and Josette in Grand Bois, Louisiana, March 9, 2012.
68 Basso (1996: 146).
69 Connor et al. (2004: 55).
70 Conversation with Theodore in Pointe-au-Chien, Louisiana, July 31, 2012.
71 Conversation with Jean in Isle de Jean Charles, Louisiana, August 2, 2012.
72 Conversation with Theresa Dardar, boat ride in Pointe-au-Chien, Louisiana, July 2, 2012.
73 Conversation with Regina in Isle de Jean Charles, Louisiana, June 1, 2012.
74 Conversation with Chris Brunet in Isle de Jean Charles, Louisiana, April 11, 2012.
75 Conversation with Chief Shirell Parfait-Dardar in Chauvin, Louisiana, May 16, 2012.
76 McNeil (2011).
77 Conversation with Chief Albert Naquin in Isle de Jean Charles, Louisiana, January 4, 2012.
78 Conversation with Victor in Isle de Jean Charles, Louisiana, July 14, 2012.
79 During Pointe-au-Chien's youth cultural camp, the children helped the elders and the Pointe-au-Chien Tribal Council and other Tribal members finish building a traditional palmetto hut, replicating the type of house their ancestors lived in until the early 1900s. Members from Isle de Jean Charles helped gather palmetto from Grand Bois, a community farther north, to make the hut because there was hardly any palmetto left surrounding the communities due to the saltwater intrusion and land loss. The palmetto hut was damaged when Hurricane Isaac hit less than two months after the hut was completed.
80 Taylor et al. (2014).

4

A LEGACY OF ATROCITIES

Establishing an Energy Sacrifice Zone

I attended the Isle de Jean Charles and Pointe-au-Chien Tribal Naming Ceremony in November 2011. About twenty of us gathered outside the Catholic Church before the service began. The tribal leaders started a small fire under the darkening sky. Tribal members from Pointe-au-Chien and Isle de Jean Charles gathered in a circle around the fire. Outside the circle a couple of male tribal members softly beat the drum, as Chief Albert entered the middle of the circle. About eight people were brought in to make an inner circle. Young adults and elders came together as smoke from the burning sage wafted around them and Chief Albert whispered in each person's ear their Indian name. The Naming Ceremony participants faced each of the four geographical directions, in turn. The slight wind caught the fringes hanging from people's clothes. Standing in the outer circle, I saw out of the corner of my eye the church congregation entering the building. As the Naming Ceremony concluded, we entered the building as well.

The Native participants being honored in the Mass entered together, two at a time walking down the aisle, the reds, turquoises, and browns of their clothes bounced slowly in rhythm to the drummers beating on one communal drum at the front of the building. The Anglo Bishop clapped at the drummers to stop and asked the congregation to honor those who were the "first settlers." I reflected on the irony of these words a few months later during the celebration for an elderly priest from the Island. The priest giving the homily speech talked about how his father had been game warden in the area for the Louisiana Department of Wildlife and Fisheries and he had known of the Island since the 1930s when his father "discovered" the Island. I thought back to the drummers playing at the front of the church, at the way their arms rose and fell together, creating one, collective beat, ensuring that their presence was undeniably heard.

Starting from the region's colonial history, this chapter focuses on how coastal Louisiana was transformed from a region of refuge into an energy sacrifice zone that risks setting the stage for future disasters. My analysis shows that government actions—from colonization to the present day—reflect a privatization paradigm that inevitably leads to energy sacrifice zones.

★★★

Story by Chris Brunet, Isle de Jean Charles[1]

Isle de Jean Charles has always been home for my family. When you take a right turn on the road here, that's my spot. I was born here, my three siblings were born here, my parents were born here. When I was growing up, we lived surrounded by family on all sides of us, aunts, uncles, grandparents.

My dad was a commercial fisherman and worked in the shipyards. As a kid in the early 1970s, I spent my days crabbing and trawling with my dad. We woke up at 4am to leave by 5am so we could be down on the lakes for sunrise. There is nothing prettier than seeing the sun come up. We would come off the boat after selling our catch at the factory and then wind down the day by taking a nap together on the floor. This was better than having toys.

When I was young, I would see armadillos, raccoons, snakes, spiders, marsh hen, otters, and some nutria. At the close of day, I would see bats flying out of the trees. I would think about the stories my elders told me of how the island was once self-sufficient, filled with cattle and gardens. Before my time things had already started to change.

We now have three large canals around us bringing water with the southeast and southwest wind. They were built for the oil and gas industry, first for boat traffic and then for pipelines. There are about five pipelines that cross this island at the upper and lower ends. The pipelines have been abandoned but the canals are still open. These man-made canals play a part in the land eroding around my house. But the oil and gas industry is not the total blame for erosion.

The change started when dams and channels were put in during the 1920s and '30s. Also, the nutrias brought here from South America ate the marsh grass and the marsh started to break up when the tide hit it. My grandma used to skin nutria and muskrats for fur, but that's no more. There is no more land left for trapping. Now, you have to leave the Island to do most anything. The saltwater is coming in and nothing is slowing down the tide.

When I was growing up, there were trees all around. This piece of wood was cut from a tree I watched grow. I'm 47 years old and I remember seeing this tree being planted. It's a Chinese tallow tree. My cousin Virgil cut down the tree a couple years ago because it was dead and rotting. The trees that once provided shade and allowed us to sit outside are all gone. The land is so saturated with saltwater that the roots can't survive. There is so much saltwater it chokes them.

Raising my niece and nephew here, they don't have as much as I had. When they get up in the morning, they don't have a chance to see armadillos, possums and raccoons in the yard, snakes on the ground, spiders in the trees. Because of the erosion, the animals have left for higher ground.

When I was their age, we had trees that blocked the west sun, so we could sit out on the porch and enjoy a long afternoon in the shade and it was comfortable. My dad played guitar and sang. He taught me how to play. We would spend the afternoon talking mostly in French and others would come by and join us. For the kids now there is some French, but more English and the trees we once sat under are gone.

Back then I could walk up the road and all alongside there was someone's house and we all knew each other. We still know each other, but now the population is scarce. Now, it's like the skeleton of the body.

I still live in the house I grew up in. My parents built it in 1960. We had high tides at that time, but the water didn't come inside. We had marsh to stop the flow of the water. Every now and then it would flood and I remember playing in the pirogue until the water went down on its own in a day or two. It has changed because the tide has eroded so much of what was here. We have got to slow the rising tide.

In 2003 my house was raised, which has allowed me to stay here. Now it is 11 feet in the air. Even though the trees have died, now I can sit in my rocker under my house and enjoy the afternoon breeze. I want to be nowhere else, doing nothing else.

My home is this, my people is this right here. It's the place I've always known as home. This land has fed our people. It may not be much, but it is ours. Being Native there is a strong connection to the land; it gives to you and you give back to it.

If a storm brought oil into our homes and the government said we couldn't go back, I couldn't put the impact into words. I still want to live here because I'm Native American, because I'm connected to the land. I'm going to live here as long as I can. I belong here.

Living here is a commitment. You have to do it in spite of the challenges of storms, flooding, distance to everything. But the good outweighs the bad. When there are no storms, no flooding, raising the kids on the Island, that is the good. It's not everywhere that you can be outside your house with a nice breeze in comfort and safety knowing everyone around you. What it was, what it is, that's what keeps me here.

FIGURE 4.1 Chris Brunet had his house raised to stay above the floodwaters, Isle de Jean Charles.

Source: Photo by author, 2012.

FIGURE 4.2 Raising children on Isle de Jean Charles, Louisiana.

Source: Photo by author, 2012.

FIGURE 4.3 Chris is an active voice for his Tribe, seen here giving an interview to the Public Broadcasting Station.

Source: Photo by author, 2012.

★★★

Colonial Legacies

The historical stage for coastal Louisiana's transformation into an energy sacrifice zone was set in 1539, when the Spanish explorer Hernando de Soto and his crew traveled through the North American southeast. Encountering a seemingly endless array of wetlands, de Soto's men described the Mississippi Valley as a "hell upon a hostile earth."[2] The arrival of the colonists created widespread and a much faster rate of environmental change throughout the southeast than changes as a result of the Native inhabitants' practices. Colonists transformed vast tracts of land into cultivated fields and introduced invasive weeds, which, along with the colonists' migration, spread and accelerated the destruction of forest ecosystems.[3] The early colonists named various landscapes and water features they came across, asserting control over history and the environment.[4]

From the 16th to 19th centuries, European colonists carried the concept of land ownership across North America,[5] converting lands once managed communally by Native people into private holdings, hindering Natives from carrying out traditional land management and adaptation strategies.[6] In 1682, the French colonist René-Robert Cavelier, Sieur de La Salle, or Robert de La Salle, traveled down the Mississippi River, claiming the surrounding land for France and naming the territory in honor of King Louis XIV. In the 1680s, La Salle established a fur trade in the Mississippi Valley and established a colony along the Gulf Coast, from which to attack Spain's silver mines in northern Mexico. The establishment of this colony ignited a series of colonial activity, with Spain working to destroy La Salle's French settlement. By 1685, the Native population in the southeast decreased from about 1.3 to 4 million to approximately 200,000 people, due largely to diseases such as smallpox introduced by European settlers.[7] Further, the introduction of new diseases, enslavement, migration of Native people from the east as British and French colonists moved westward, replacement of traditional hunting and gathering grounds with agricultural development, and colonial military actions and rivalries contributed to changing tribal social structures of Native inhabitants.[8] As a result of so many Natives dying in a very short period of time, much knowledge of history, traditions, and traditional medicine was lost.[9]

In the early 1700s, the major port of New Orleans made colonial Louisiana an important trade center, pulling the region and its inhabitants into a global market. By 1720, under the auspices of the Company of the Indies, a French corporation, French settlements emerged along the Gulf Coast and Mississippi River. Encouraging slave importation from West Africa and the Caribbean's French Islands to entice economic development, between 1719 and 1731 the Company brought over nearly 6,000 African people to work the larger concessions, otherwise known as plantations, as slave labor.[10]

The Company's marketing of Louisiana created one of the first economic bubbles, known as the "Mississippi Bubble." While thousands of investors in France bought shares in the Company, thought was not given to financing the development of Louisiana. Thus, as expenses in Louisiana increased, the Company could not pay dividends on all the stock it sold, causing the downfall of the Royal Bank in France and the burst of the "Mississippi Bubble."[11]

During this time, one of the most important crops that the French cultivated to shape the landscape and culture, and that a number of slave laborers already knew how to plant, was rice, which sustained the people, supported the creation and expansion of export trade in sugar, cotton, tobacco, and indigo, and convinced the royalty and elite of France to invest in Louisiana. The French turned the lower Mississippi Valley into dry land for rice fields and initiated the transformation of French Louisiana from a colony based on trade with the Native population into a colony based on agriculture.[12]

Separating Land and Water

The French settler Joseph Villars Dubreuil was the first person to put slaves to work on a levee system to drain land for agricultural development. Colonists soon considered such a system fundamental for productive agricultural development in the lower Mississippi Valley; separating land and water gave the French more control over both. Whereas landowners first built houses on high ground fronting the river, they moved the houses back to make room for levees, placing the houses into lower lying areas and at greater risk of inundation, necessitating still higher levees.[13]

During the winter, slaves on the plantations were put to work building levees and cutting cypress to turn into lumber. Deforestation increased rain runoff, reducing flooding where land was cleared but increasing flooding downstream. Such a process demanded more tree clearing and more levees and ditches. French and Spanish colonial rule made levee construction a legal indication of land ownership.[14]

The different colonial rulers had differing ideas about water management. Spanish colonists debated whether or not to interfere with the flow of the Mississippi River. French engineers believed in exploiting natural river courses without really altering them, whereas engineers working for the United States followed the British tradition that held canals were superior to rivers because they were easier to control.[15] The US ideal of river management won after the French sold Louisiana to the United States in the 1803 Louisiana Purchase.

Absorption into the United States

Driven by a desire to expand American trade, in the early 1800s US President Thomas Jefferson decided that the United States needed to find a way to take control of New Orleans, which served as an important port for commerce. The United States negotiated a treaty with France and bought the Louisiana territory for $15 million. The Purchase joined New Orleans with lands west of the Mississippi River and placed it under US control.[16] Louisiana became a state in 1812. The first Louisiana constitution created a state government designed to keep the framers of the state constitution and those similar to them in power: only White, male landowners could aspire to office, and only those who paid taxes could vote.[17]

The American colonists concluded the process of drying the lower Mississippi Valley for cotton plantations nearly 200 years after the French started the process for rice fields.[18] The increasing value in the market for cotton and expansion of transportation and new technologies put further demands on the acquisition of Indian lands in the southeast. Pursuing ever farther westward American expansion, Secretary of War Henry Knox promoted the transformation of Native people's economies into intensive agriculture and private land ownership. President Jefferson continued the federal program put in place under President George Washington to "civilize" Native people; government trading posts were set up to persuade Native people's dependence on American goods, encouraging them to take on debts that could be paid off by granting lands to the United States.[19] Lands that were once held communally came under private landholdings.

Jefferson's policies and strategies toward Natives led to nearly a dozen tribes ceding approximately 200,000 square miles of land, leaving Natives with two choices: forced assimilation or removal. The Louisiana Purchase provided the means for Jefferson's administration to remove Natives from east of the Mississippi River to the west. Jefferson's strategy paved the way for President Andrew Jackson's formal national Indian removal policy.[20]

An Era of Indian Removal

Before Andrew Jackson was elected US President in 1829, Indian removal was negotiated between the federal government and tribal authorities. However, once Jackson came into power, states began to pass laws abolishing tribal existence and extending the state's jurisdiction.[21] Leading up to the passing of the Indian Removal Act of 1830, President Jackson argued for the Act's passage based on negotiating state and government affairs, westward expansion, and framing Indians as inferior to the Anglo race:

> [The act] puts an end to all possible danger of collision between the authorities of the General and State Governments, on account of the Indians. It will place a dense and civilized population in large tracts of country now occupied by a few savage hunters. By opening the whole territory between Tennessee on the north, and Louisiana on the south, to the 'settlement of the whites, it will incalculably strengthen the southwestern frontier, and render the adjacent States strong enough to repel future invasion without remote aid. It will relieve the whole State of Mississippi, and the western part of Alabama, of

Indian occupancy, and enable those States to advance rapidly in population, wealth, and power. It will . . . perhaps cause [Indians] gradually, under the protection of the Government, and through the influence of good counsels, to cast off their savage habits, and become an interesting, civilized, and Christian community.[22]

On May 28, 1830, the US Congress passed the Indian Removal Act, "[a]n act to provide for an exchange of lands with the Indians residing in any of the states or territories, and for their removal west of the river Mississippi." Subsequently, tens of thousands of Native people—forced west out of their ancestral homelands along the Eastern seaboard and into set land borders in tenuous new territories—lost their lives along this journey, dying along the way from diseases such as tuberculosis, dehydration, and other adversities.[23] An estimated 6,000 Cherokee alone died along the approximately 1,200-mile migration route, dubbed the "Trail of Tears."[24] For those who survived, the government-dictated removal and reservation boundary limits led to intense loss of cultural identity and sense of place.[25] The Choctaw described their new lands in the west as "the Land of Death."[26]

The Tribes moved across the Mississippi River to lands in the west that were part of the Louisiana Purchase. During this time, Louisiana's densely forested high ridges at the southern end of the bayous served as a "region of refuge,"[27] or ecological shelter, for some Native people from the southeast. Individuals and families that avoided assimilation were ignored if they fled to lands of marginal value, such as the swamps.[28] During this time, the ancestors of Isle de Jean Charles, Grand Caillou/Dulac, and Pointe-au-Chien Tribal members fled south down the bayous into present-day Terrebonne and Lafourche Parishes. As François, an elder from Grand Caillou/Dulac, shared, "My grandma remember the Trail of Tears where they picked up all the Indians and put them on reservation. My grandmother told me they moved into swamp because didn't want to be taken away. Explorers weren't brave enough to go into swamps."[29] During their journey, they intermixed with the French and Cajuns (Acadians) that had recently arrived in the region.[30]

Establishing Settlements

The Cajuns originated during the sixteenth and seventeenth centuries in the colony of Acadia in French-controlled Canada. The colony was placed under British control in 1713, by which time the Acadians had become a culturally distinct, French-speaking people. The Acadians' diaspora, known as the *Grand Dérangement*, was largely the consequence of the Acadians' refusal to become part of the Anglo-French imperial wars in North America during the 18th century. Starting in 1755, many Acadians were forcibly removed by the British from the French colony of Nova Scotia and sent to France, other British colonies, and the West Indies. Thousands of Acadians arrived in Louisiana during the 1770s and 1780s. By the latter part of the 1700s, the Acadians had developed into the predominant cultural group in south Louisiana.[31] Most Acadians settled west of the Mississippi River in the bayou areas along the southwestern prairie, developing a rural lifestyle based on farming and hunting.[32]

The tribal residents, including those who intermixed with the Cajuns, established settlements in the previously uninhabited southern ends of the bayous. Once people escaped down the bayous, they were physically and socially isolated from communities and political relations farther north. Families lived in small clusters and maintained a subsistence culture based on fishing, trapping, hunting, and farming. They had a wealth of resources, including barrier islands, extensive, life-giving estuaries, and an abundance of aquatic resources.

The communities remained isolated until the early 1900s when missionaries started visiting the southern ends of the bayous. For example, according to an account recorded by journalist Gérard Pelletier, a reverend in Louisiana told of his first encounter with the area in the 1940s, speaking with the owner of a local general store:

I do not imagine that anyone lives there, in . . .

To the contrary, Father. . . . You would be surprised to see how many families live out there . . .

But, how can people live in the swamps? How can I go to them?

Oh! You do not have to worry about them. They are Sabines, you know.

Sabines?

Yes, this is the nickname for the Indians around here.[33]

The institutionalized racism and segregation in Louisiana that the dialogue above points to increased each tribal settlement's social cohesion and resulted in residents having minimal contact with outsiders.

Muddling Bi-racial Segregation Laws

In 1890, the first official segregation law, commonly known as Jim Crow, was passed in Louisiana that required all railroads carrying passengers to provide "separate but equal" accommodations for White and "Colored" people.[34] The presence of Indians as another racial group in the region muddled the bi-racial Jim Crow laws. Local Whites and officials in Louisiana, as well as federal government officials, racialized Indians as "colored," diminishing tribal rights and creating challenges for Indians to establish their place in a society that labeled itself as bi-racial.[35] Theresa, from Pointe-au-Chien, described how the bi-racializing was emphasized in everyday practices and even from the very first moment of declaring one's identity: most hospitals in the area at the time provided only two racial options for people to check on a birth certificate—White or Black.

In a complex web of colonialism, political status, and racial structures, structuring society bi-racially forced a wedge between Indians and Blacks in the southern United States, including Louisiana.[36] For the state and local governments wanting to take over Indian lands, they tried to label Indians as "Colored" in large part to erase their presence entirely and further reduce their socioeconomic status. This led to Indians creating more social distance between themselves and Blacks to try to retain their lands. By keeping Indians and Blacks from coming together, the White minority could retain power.[37]

In the mid-1900s, signs appeared in Terrebonne Parish saying "No Indians," "No Colored Allowed," and "Whites Only."[38] The tribal residents were racially segregated in schools and churches until the 1960s. For those who grew up before the 1970s, some interacted with extended family members living in other tribal communities but rarely interacted with anyone else. But as the landscape changed, so too did the communities: more and more outside actors (e.g., oil corporations and private land developers) came in, more locals left, and the southern end of the bayous were no longer politically or socially isolated. The territories the Tribes settled and waters they fished started being taken over and transferred into others' domains. The story was all too familiar.

Controlling the Ecosystem

Following the end of the US Civil War in 1865, most of the land in the lower Mississippi Valley came into the possession of colonial descendants. The region's river management system began with the draining of the wetlands for agricultural production, then expanded with the construction of levees for flood control. The new landowners succeeded in obtaining congressional support for federal flood control by defining levees as being built primarily to support navigation.[39] Control of the Mississippi River was promoted to be in the nation's interest for the flow of commerce and westward expansion.

The flood of 1849 led to debate over whether or not to control the Mississippi River. When the Mississippi Delta flooded beneath three feet of water in 1858, the question was no longer *whether* to control the river but *how* to control it.[40] The federal government enacted legislation focused on flood control in the Mississippi Valley. The Swamp Lands Acts of 1849, 1850, and 1860 gave the federal government control to transfer unsold swampland[41] to state governments. The state could then sell the land to private interests, with the condition that revenue from the sale of the land be used to fund flood control measures. Thus, the federal and State of Louisiana governments entered the Industrial Age with new policies and technologies to further the interests of private landowners and political elite.

Louisiana and the Age of Industrialization

As Louisiana entered the Industrial Era, the Great Mississippi Flood of 1927, one of the worst floods and greatest disasters in US history, inundated over 20,000 square miles of land in the Mississippi Valley and displaced over 600,000 people.[42] Thereafter Congress appropriated funds for levee reconstruction and the Mississippi River and Tributaries Project.[43] The Flood Control Act of 1928 gave the federal government control to seize lands needed for easements and rights of way, providing landowners with compensation in return.[44] Thus, private property was taken by the federal government and developed to protect national interests.

Following the 1927 flood, the Mississippi River Commission, US Army Corps of Engineers (USACE), and the federal government pursued a river management strategy that kept the water out, the land dry, and the Black laborers within its boundaries.[45] Flood protection levees were built, and dam and reservoir construction on major tributaries occurred along the Mississippi River, along with channels and navigation canals. The levees worked to segregate land and water, as well as further separating Whites and Blacks based on ownership of dry land, perpetuating Jim Crow policies.

Kerry St. Pé, the Executive Director of the Barataria Terrebonne National Estuary Program (BTNEP), described how, following the 1927 Flood,

> People blame the Corps of Engineers for building levees but they forget at the time the people demanded the levees be built. Big corporations, big land corporations, agricultural corporations wanted their land protected. So it got protected. There was pressure from the major land corporations. It was an easy sell. Yeah, we gonna protect y'all from flooding. It's like how every environmental decision is made even today. Pressure from the big corporations, the rich.[46]

Scientists raised issues about this kind of river management system: they were concerned that this approach viewed water piecemeal instead of as a whole system. In 1928, Percy Viosca Jr., a biologist, wrote, "Man, interfering with nature, has created new conditions of existence in our wet areas, and . . . further decline is inevitable unless some effort is made to restore the former state."

The 1930s saw a continuation of the flood control measures enacted in the 1920s around the country. In response to the Great Depression, President Roosevelt's New Deal focused on economic recovery and job creation, including the establishment of a large public works program, and the construction of dam and reservoir projects throughout the country. Most significantly, the Tennessee Valley Authority, the federal government's most widespread and comprehensive river valley development program, was enacted in 1933, which authorized the construction of "dams, and reservoirs, in the Tennessee River and its tributaries . . . [to] control destructive flood waters in the Tennessee and Mississippi River drainage basins."[47]

Aiming to control the Tennessee River, the federal government, through the Tennessee Valley Authority, purchased vast amounts of land under eminent domain to construct dams and reservoirs, displacing thousands of families, including subsistence farmers and their tenants.[48] The Tennessee Valley Authority's model of controlling waterways to form a productive landscape for economic benefits jumpstarted the post–World War II global interest in constructing large dams,[49] as well as serving as the model for controlling waterways in the Mississippi Delta.

While the river management system put in place in the twentieth century by the USACE and local levee districts provided both flood control and economic benefits, such forms of management, control, and re-direction of the Mississippi River deprived the coastal system of much needed sediment and fresh water.[50] Instead of collecting along the coast, the sediment brought by the Mississippi River went into the Gulf of Mexico and dropped to the bottom of the sea floor.

During this time, the vast expansion of nutrias throughout coastal Louisiana also greatly affected the wetlands. Nutrias, which are large rodents, were brought to the United States from South America in the late 1800s for fur ranching. The market for nutria fur was strong through the 1960s and 1970s. However, by the end of the 1980s, the foreign and domestic fur markets declined, and there were reports of nutrias causing substantial damage to marsh and agricultural lands. In 1999, a Louisiana coast-wide nutria damage assessment estimated that nutrias had damaged 105,000 acres of marsh.[51]

Louisiana also marked the beginning of the Industrial Era through exploitation of natural resources, including petroleum, natural gas, sulfur, salt, and timber. By 1888, after Congress expanded opportunities for the transfer of public lands, more than 1.7 million acres of Louisiana timberland belonged to 41 northern lumber companies that had already exhausted the forests of the Great Lakes region.[52] Larger companies typically moved on after clear-cutting the land, devastating forests with no thought to reforestation. Drying the land and deforestation meant the need for fertilizers to supplement the once rich soil, and minimal barriers to insects in the lower Mississippi Valley. This resulted in farmers using chemical pesticides, which flowed into the Mississippi River and contaminated the water and inhabitants downstream.[53] Contemporarily, runoff containing highly contaminated nitrogen and synthetic chemicals from large-scale agricultural development continually flows down the Mississippi River into the Gulf of Mexico, creating the largest hypoxic "dead zone" in US waters and damaging fisheries.[54]

Combined with the impacts from the levees and agricultural drainage, by 1928, 85 percent of the 14.5 million acres of upland forests were cut, resulting in more rapid runoff and worse droughts and floods. In addition to the upland deforestation, canals were cut to remove and sell the timber from swamps and to aid navigation, and channels were deepened for drainage, resulting in erosion of the natural ridges and coastal land.[55] With coastal Louisiana's economic activities already dominated by extractive industries like logging, the timber industry set the precedent for industry/government cooperation.[56]

In parallel, in 1901, 30 miles from the southwest corner of Louisiana, in Beaumont, Texas, the first oil gush occurred, marking the beginning of the modern petroleum industry in the United States. Immediately following, oil exploration began in Louisiana with the state's first oil field discovered in 1901 near Jennings in southwest Louisiana. In 1916, a large natural gas field was discovered near Monroe in northern Louisiana. By 1922, Louisiana's annual petroleum production made up 8 percent of the total production in the United States.[57]

After the end of World War II in 1945, there was an immense growth of national energy consumption. An increase in automobiles, urban sprawl into suburbs, highway construction, and low prices for oil and gas dramatically increased the demand for fossil fuels. In the 1950s, the national demand for energy contributed to the creation of a vast petrochemical industrial complex along the Mississippi River between Baton Rouge and New Orleans—now commonly referred to as "cancer alley"—an area that historically includes predominantly African-American communities.

The national energy demand also led to the construction of hundreds of oil and gas platforms along the Louisiana coast in order to pump out crude oil and natural gas from the earth. With a seeming endless supply of natural resources, the state government allowed industrial development and permitted highly toxic chemical waste to be dumped into the Mississippi River.[58] After centuries of environmental transformation and resource exploitation, by the middle of the 20th century, control over oil resources emerged as the dominant form of power in Louisiana and throughout the world.

The population in the area rose drastically in the middle of the 20th century during the economic boom due mostly to oil extraction, as well as the expansion of the Intracoastal Waterway and the Houma Navigation Canal. Terrebonne Parish doubled in population size from 1930 to 1960—increasing from 29,816 to 60,771 people.[59] The discovery of oil attracted outsiders to the bayous, turning the tribal communities' isolation into a thing of the past. Multinational corporations, along with recreational fishers and tourists, moved in, intruding on the close-knit tribal fishing communities' lifestyle and culture; residents now had to compete for natural resources.[60]

Living in an Energy Sacrifice Zone

Like the region's river management system—initiated by settler colonists who drained the land for agricultural production and built levee systems to control flooding—the oil and gas industry dredged passageways through the marsh for pipelines and navigation, once again extending human control over the environment. This was made possible, in part, by the passing of the Mineral Policy Act of 1920, which authorized the federal government to offer a ten-year lease of federally owned lands to private individuals and companies to extract petroleum and other minerals.[61]

Starting with the first coastal zone oil lease in 1921, the dredged passageways enabled water to rush in from the Gulf of Mexico during storms and high tides, causing severe land loss, erosion, and saltwater intrusion (Figure 4.4).[62] The colonial

FIGURE 4.4 Passageways cut through the marsh by oil and gas corporations for pipelines and boundaries near the three communities.

Source: Photo courtesy of Babs Bagwell, 2012.

settler process continued into modern times; by the 1980s, five private corporations owned one-quarter of the entire Louisiana coast.[63] There are now approximately 25,000 miles of pipelines and 3,500 offshore production facilities in the central and western Gulf of Mexico's federal waters, three-quarters of which are off the coast of Louisiana.[64]

To grasp the scale of the changing landscape, it is helpful to see it on the ground through the eyes of local residents and hear their stories, and also see it from above. After several months of listening to grounded stories, I did a flyover of the area to see the broader view. Taking off from Houma and flying south, I saw the point where the remaining forest of trees transitioned into fields of skeletal remains. I kept thinking to myself, there is no such thing as a straight line in nature. But that was what caught my eye in every direction—the straight lines cut for canals that overwhelmed the water- and landscape.

Standing on the ground, one can see remnants and continued operations of the oil industry all around. For example, sitting together on the deck of his elevated house (about 14 feet off the ground), Chris (Isle de Jean Charles) pointed to the abandoned wells surrounding the Island and where land and trees had once stood. He told me about the noise and light pollution that became part of the land-scape growing up: "Back along this back bayou, there was three wells that I know of. It was so close to the house you could see the derricks and lights and whenever they'd start drilling you could hear it all. That's [well] number four behind that house. And also had some on my uncle's property, first well on the Island." He described how the oil companies would "make a big old opening where we used to have marsh and just leave it as it was."[65]

As the oil industry became further embedded in the physical topography and economic diversification declined, the industry, law professor Oliver Houck (2015: 192) explained, "not only molded the politics and economics of Louisiana, it molded the mind." The majority of jobs now available locally are with the oil and gas and supportive industries, which has moral implications for inhabitants who have to then work for the very industry that is responsible for the environmental degradation of their land and livelihood loss. While people often did better economically working for the oil industry, they had to reconcile being fishers, as Michele (Pointe-au-Chien) described, "in people's hearts," and no longer having much time for fishing because of other jobs (Figure 4.5).[66]

Many local residents—mostly males—now work on oil rigs or tugboats, moving vessels through canals, often for the oil industry, and some work as welders for the oil and shipping industries. While providing income, the oil industry jobs tend to change family and community dynamics. Driving tugboats allows the men to still be out on the water. However, while many of them used to be gone for a few days at a time shrimping, the trips as tugboat captains are longer, often lasting for three weeks at a time, and then returning home for a week before leaving again. Or people had to move to find work elsewhere, which also broke apart families and changed dynamics within the community.

FIGURE 4.5 The Tribes have experienced severe subsistence and livelihood impacts. Isle de Jean Charles, Louisiana.

Source: Photo by author, 2012.

The state's reliance on one type of resource production and global market integration created a fixed economy that hindered economic expansion and new job creation. It is a situation reminiscent of an Appalachian coal industry that, anthropologist Bryan McNeil (2011: 65) noted, systematically prevents residents "from developing community resources in ways outside the state's agenda—an agenda that systematically protects coal." In Louisiana, the growing connection between the government and oil and gas industry set the stage for new disasters to unfold; it did so primarily by turning the region into an energy sacrifice zone, what geographer Geoffrey Buckley and Laura Allen (2011: 171) defined as "a place where human lives are valued less than the natural resources that can be extracted from the region."

The Oil and Gas Industry/Government Relationship

The oil and gas industry/government relationship that has helped produce the decimation of coastal Louisiana is reflected in several decades of policy and regulation. At the national level, the US Department of Interior has historically adopted practices and standards developed by the American Petroleum Institute (API) as formal regulations.[67] Every major oil company Chief Executive Officer (CEO) is on the American Petroleum Institute board.[68] This raises questions as to whether these practices and standards make operations safer or, alternatively, encourage industry sovereignty and profitability by eliminating government oversight.

Kerry St. Pé, the Executive Director of BTNEP, further elucidated the government-corporate relationship:

> Most people blame the oil and gas companies for all the canal digging. But I don't put as much blame on them as I put the blame on the federal and state government, principally the state government because most of those canals were dug with permits, they were permitted. And I believe corporations are just like children. If they ask for the candy and you give them the candy they're going to take it. They ask for a canal, you give them a canal, (snaps fingers) they're gonna dredge it.[69]

The oil and gas industry is exempt from significant provisions of seven major federal environmental laws including the Safe Drinking Water Act; Clean Air Act; Clean Water Act; Resource Conservation and Recovery Act; Comprehensive Environmental Response, Compensation, and Liability Act; National Environmental Policy Act; and Toxic Release Inventory of the Emergency Planning and Community-Rights-to-Know Act.[70] One example is that storm water discharges from oil and gas drilling and production activities are exempted from the Clean Water Act's permitting requirement for all discharges of pollutants to rivers, streams, creeks, and wetlands.[71] Furthermore, the Deep Water Royalty Relief Act of 1995 cut the already low fees the United States was charging oil companies to drill on the Outer Continental Shelf.

Specific to Louisiana and the Gulf Coast, the 1978 Amendment to the Outer Continental Shelf Lands Act of 1953 singled out the Gulf of Mexico for milder environmental oversight, exempting oil and gas lease-holders on the Outer Continental Shelf[72] from submitting development and production plans, including environmental safeguards, for agency approval.[73] The State of Louisiana also provides corporations with over $1.79 billion per year in subsidies, incentives, and tax breaks, with a large portion of this money going to the oil industry.[74]

A number of Louisiana government officials are directly connected to the oil and gas industry. Some examples include: Louisiana State Senator Robert Adley previously owned Pelican Gas Management, Inc., and served as a board member for the Louisiana Oil and Gas Association.[75] Louisiana State Representative Jim Morris was a former oil executive.[76] Chris John served in the Louisiana Congress, including serving on the House Natural Resources Committee, and US Congress before becoming president of Louisiana's Mid-Continent Oil and Gas Association (LMOGA), a trade association representing the oil and gas industry operation in Louisiana and the Gulf of Mexico.[77]

Louisiana's major environmental agencies also have deep connections to the industry. For example, Jim Porter served as Louisiana's Secretary of Department of Natural Resources in the 1980s before becoming president of LMOGA.[78] Subsequently, the first Deputy Secretary of Louisiana's Department of Environmental Quality went on to become president and chief lobbyist of LMOGA.[79] J. P. Batchelor, a former executive of Amoco, became head of Louisiana's Department of Natural Resources' Office of Conservation, and Scott Angelle started a coalition of oil and gas groups to oppose new federal regulations while he was Secretary of Louisiana's Department of Natural Resources.[80] Oil interests also highly influence state-promoted restoration efforts. For instance, America's Wetland Foundation— a "balanced" forum established for problem solving about coastal land loss in Louisiana—is led by Shell Oil Company as a primary sponsor,[81] as well as Chevron, ConocoPhillips, and ExxonMobil as its sustainability sponsors.

The relationship between the oil industry and the state was further highlighted when the Southeast Louisiana Flood Protection Authority-East (SLFPA-E) filed a lawsuit in July 2013 against 97 oil, gas, and pipeline companies for "ravag[ing] Louisiana's coastal landscape," demanding that the companies restore the damaged wetlands or pay for irreparable damages.[82] Only three SLFPA-E board members voted against the resolution supporting the lawsuit, all of them appointed by Louisiana Governor Bobby Jindal.[83] Claiming it was not in line with Louisiana's coastal restoration policy, Jindal subsequently signed legislation blocking the lawsuit and preventing Louisiana government agencies from accepting litigation challenging the oil and gas industry. Local environmental groups claimed that Jindal's opposition was tied to the over $1 million he received in political contributions from the oil and gas industry.[84] Seventeen additional parishes have since joined together to file a lawsuit against the oil and gas companies.

In March 2017 a federal appeals court, upholding a 2015 decision, denied reviving the SLFPA-E lawsuit, resolving the lawsuit concerned enforcement of federal laws, and should therefore not be tried in a state court, which is where the levee authority wanted the case to be tried. The court also agreed that the SLFPA-E did not prove that the oil and gas companies were obligated to restore any damages caused by their operations and that the state of Louisiana does not require the companies to restore the land to its "natural state."[85] While the lawsuits continue to languish in the court system, their outcomes could have subsequent effects on other lawsuits against oil and gas corporations, including claims against BP for compensation following the 2010 BP Deepwater Horizon Oil Disaster.

Manufacturing a Sacrifice Zone: Accumulation by Dispossession

Neoliberal policies emphasizing free trade and privatization encourage the idea of a sacrifice zone.[86] In the case of the oil and gas global production network, neoliberal capitalist policies have resulted in what geographer Michael Watts (2012: 458) described as "frontier dispossession and reckless accumulation," in which oil states support petro-capitalism and the logic of oil extraction is a central component in "the making and breaking of community."[87]

In coastal Louisiana, once land disappears under water, the state takes possession and can lease the submerged area to oil and gas corporations. Laws such as the Louisiana Civil Code 450 enable this process: "Public things that belong to the state are such as running waters, the waters and bottoms of natural navigable water bodies, the territorial sea, and the seashore."[88] The state is allowed to claim the minerals under the water as well.[89] These leasing and ownership schemes affect communities' restoration efforts. Often, the Tribes need to initiate restoration activities in the water surrounding significant places, such as ancestral mounds, but cannot do so because the surrounding water, which had once been land, was taken over by the state and leased to oil and gas corporations.

Land grabbing is not a new process in Louisiana, having occurred since the onset of colonization. In the 1950s, a number of residents left the southern parts of the bayous in Terrebonne and Lafourche Parishes as land developers and corporations settled in the area. Individuals and private companies could use the colonial-developed court system to force individuals to sell or lease their lands. Groups of people who held land communally did not have a government-recognized legal title, enabling outsiders to seize the land.[90] Pointe-au-Chien's Chairman Chuckie Verdin explained that because people could not read the forms they were given by developers or corporate representatives about their land, they would "just put a cross. So someone wants to forge a cross, they can. Was told someone signed a form after he was dead for his property."[91] Taking advantage of people's illiteracy, some operators told people they were only signing lease agreements when they were really selling their land.[92]

This happened all over coastal Louisiana. For example, Victor, an Isle de Jean Charles Tribal member that had relocated to Mississippi, described how his family in Lafitte, north east of the three tribal communities, had their property taken by the Louisiana Land and Exploration Company.[93] After land and oil developers arrived, instead of property being passed down from one generation to the next, residents needed official papers and documents to prove property ownership. Such policies play a major role in creating what geographer David Harvey (2003) dubbed "accumulation by dispossession,"[94] which includes the commoditization and privatization of land and involves the appropriation of the non-human environment by the wealthy, ruling class, highlighting the loss of environmental, as well as social and economic rights. Further, with purchasing power and control over local resources, multinational oil corporations and private land developers contemporarily purchase vast quantities of land on higher ground just north of the communities, leaving the Tribes and other coastal communities with few options as their lands become further inundated with saltwater.

Corporatization, monopolization, and consolidation perpetuate the land grabs and difficulty in restoration efforts and holding any one entity accountable. Trying to parse out when specific oil and gas companies came into the area and where they built pipelines and wells, I asked some Isle de Jean Charles Tribal members during a story circle at Pierre and Marie's house who was the first oil company to come into the area:

Maurice: Texaco.
Louis: Oh yeah Texaco was the first one, Texaco, Exxon, and Shell. I know in Pointe-au-Chien it was Texaco.
Pierre: No, Humble.
Louis: Yeah, Humble.
Pierre: Texaco was Lake Barre.
Louis: Oh yeah, Golden Meadow.
Maurice: Leeville.
Pierre: This Island it was the Humble. And the other name, Esso.
Maurice: It's Exxon now.[95]

The above passage illustrates the confusion brought by the different companies coming in and out of the area through the 20th century, consolidating, changing names, and buying out other companies. For example, the Louisiana Land and Exploration Company, a Maryland-incorporated corporation based in Texas, covered over half of Louisiana's two million acres available for oil exploration by 1928. At that time it formed an agreement with the Texas Company, which became Texaco, guaranteeing it the right to explore for and produce oil and gas on Louisiana Land's properties in Louisiana.[96] Another example is Humble Oil, which consolidated their US operations with Standard Oil in the late 1950s. Humble then took over Esso in 1960, and in the 1970s Humble became Exxon; however, the products were marketed under these different names in different places.[97]

British Petroleum, highlighted in the next chapter's discussion about the BP Deepwater Horizon Oil Disaster, has been a major player in the monopolization and consolidation trend. During World War I, the British government appropriated BP's assets, which at the time was a German-owned company marketing its products in Britain.[98] These assets were then sold to Anglo-Persian, a company initiated in 1908 when Britain acquired Persian Oil and renamed it Anglo-Persian. The name switched again to Anglo-Iranian Oil Company in 1935, and in 1954 it took the name British Petroleum Company. The British government sold the last of its shares in the company in 1987. At that time, BP bought Standard Oil of Ohio, and in 1998 bought Amoco, or the American Oil Company, followed two years later by acquiring ARCO, or the Atlantic Richfield Company, all of which were marketed separately but owned by BP.[99]

Claiming the Commons

The Tribes claimed ownership of property before the oil corporations and developers arrived and land was traded among the Tribes' ancestors in the area,[100] but land and water use was also managed as open-access. With local knowledge of the environmental and resource conditions, people who depend on the resource for their livelihood are more likely to understand that monitoring the use of the resource provides long-term benefit.[101] The commons are honored by people whose livelihoods, way of life, and cultural identity are formed through common resources such as land and water.[102]

The commons, as defined by anthropologist Donald Nonini (2007: 1), is "those assemblages and ensembles of resources which human beings hold in common or in trust to use on behalf of themselves, other living human beings and past and future generations of human beings, and which are essential to their biological, cultural, and social reproduction." The commons approach includes "humans as active participants in the environment."[103] A major flaw with private ownership of the commons is that for resources to not be exploited, political economist Elinor Ostrom et al. (1999: 281) noted, "users must be interested in the sustainability of the particular resource so that expected joint benefits will outweigh current costs."

There is an inherent conflict of interest when different actors that do not share a common vision of the resources control the waters and lands surrounding the communities through the laws of privatization and free market enterprise.[104] For example, the Chief Executive Office (CEO) of the Apache Corporation,[105] which owns land around the communities and is one of the world's biggest oil and gas exploration and production companies, stated, "Since its inception in 1954, Apache has been driven by a relentless pursuit of opportunity to profitably grow an independent oil and gas company for the long-term benefit of our shareholders."[106] Apache produces oil and natural gas on five continents and anyone can buy shares of Apache stock.[107] The shareholders are located all over the world, removed from the local landscape and continued sustainability of the lands and waters. A conflict occurs when the extraction of one thing (such as oil) for some people's benefit causes the deterioration and loss of other things (such as land and aquatic species) for other people's well-being.

Residents do receive some royalties from the oil extraction. Chris (Isle de Jean Charles) noted, "[Oil companies] made so many promises . . . if they didn't find nothing you didn't get nothing, and if they did they took it and you didn't get nothing. Until this well started producing a little bit . . . My first royalty off of that was $9.35."[108]

★★★

In June 2012, I stood in a room of the Superdome in New Orleans a few feet away from US Secretary of Interior Ken Salazar while he gave a press conference. He had just opened the bids for oil leases off coastal Louisiana. I looked out at the sea of hundreds of mostly White, gray-haired men. The bids on dozens of leases ranged from hundreds of thousands of dollars to multi-millions, like BP's bid at $27 million. I walked out of the room and looked down at the floor of the Superdome, trying to imagine this place housing thousands of survivors in the days following Hurricane Katrina seven years earlier. I could still hear them reading off the bids. It seemed that not much had changed since Katrina, or since the 2010 BP Deepwater Horizon Oil Disaster, the largest oil spill thus far in US history, which further exposed the role of government and the oil and gas industry in manufacturing the energy sacrifice zone.

Notes

1 Digital story, audio recorded on July 2 and July 27, 2012.
2 Kane (1944: 5).
3 Gremillion (2004: 67).
4 Morris (2012: 39–40).
5 The land ownership concept is first denoted in the Inter Cetera papal bull of 1493, which granted to Spain "the right to conquer the lands which Columbus had already found, as well as any lands which Spain might 'discover' in the future" (Newcomb, 1992).
6 Redsteer et al. (2010); Redsteer et al. (forthcoming).
7 Saunt (2004).
8 Cummins (2014: 16); Williams (1979b: 14).
9 Saunt (2004: 128).
10 Morris (2012); Cummins (2014).
11 Cummins (2014: 51).
12 Morris (2012).
13 Ibid.
14 Ibid.
15 Ibid.
16 Cummins (2014).
17 Schafer (2014: 125).
18 Morris (2012).
19 Williams (1979b); Dowd (2004).
20 Dowd (2004); Hirsch (2009).
21 Dowd (2004).
22 Jackson (1830).
23 Blackburn (2012).
24 Native Voices (2015).
25 Bartrop (2007: 184–185); also Blackburn (2012).

26 Akers (1999: 73).
27 Vélez-Ibáñez (2004).
28 Williams (1979a: 198).
29 Conversation with François in Bush, Louisiana, May 31, 2012.
30 Terrebonne Genealogical Society (1998); Westerman (2002).
31 Brasseaux (1985, 1991).
32 Cummins (2014: 86–87).
33 Pelletier (1972: 8).
34 Haas (2014: 252).
35 Klopotek (2011).
36 Klopotek (2011); Perdue (2012).
37 Williams (1979a: 198–202).
38 Truehill (1978).
39 Barry (1997).
40 Morris (2012: 140–154).
41 Swamps in Louisiana are forested wetlands (EPA, 2013).
42 Barry (1997).
43 Morris (2012: 166).
44 Barry (1997).
45 Morris (2012).
46 Conversation with Kerry St. Pé in Thibodaux, Louisiana, July 27, 2012.
47 Tennessee Valley Authority (1961: 4).
48 McDonald and Muldowny (1982).
49 D'Souza (2008).
50 Barry (1997); Turner (1997); Laska et al. (2005); Freudenberg et al. (2009); CPRA (2012).
51 Holm Jr. et al. (2011).
52 Haas (2014: 315).
53 Morris (2012: 183).
54 Coastal Louisiana Ecosystem Assessment and Restoration/CLEAR (2006); CPRA (2012).
55 Viosca Jr. (1928: 229).
56 Freudenberg and Gramling (2011: 135–136).
57 Haas (2014: 317–318).
58 Kurtz (2014: 369).
59 Family Search (2014).
60 Solet (2006).
61 Freudenberg and Gramling (2011: 87).
62 Turner (1997); Austin (2006).
63 Houck (2015).
64 Freudenberg and Gramling (2011: 171).
65 Conversation with Chris Brunet in Isle de Jean Charles, Louisiana, April 11, 2012.
66 Conversation with Michele in Pointe-aux-Chenes, Louisiana, December 20, 2012.
67 National Commission on the BP Deepwater Horizon Oil Spill and Offshore Drilling (2011).
68 Juhasz (2011: 282).
69 Conversation with Kerry St. Pé in Thibodaux, Louisiana, July 27, 2012.
70 Environmental Defense Center (2011).
71 Ibid.
72 The Outer Continental Shelf is a landmass extending out from the coast under shallow waters, with "outer" referring to the lands that are more than three miles offshore and under federal jurisdiction (Freudenburg and Gramling, 2011: 101).
73 National Commission (2011).
74 Silverstein (2013: 48–49).
75 Louisiana State Senate (2014).

76 Silverstein (2013: 54).
77 LMOGA (2014).
78 Gill (1989).
79 Houck (2015).
80 Silverstein (2013: 54).
81 Burley (2010: 118–119); Houck (2015).
82 Jones et al. (2013: 3).
83 Schleifstein (2013a).
84 Schleifstein (2013b).
85 Schleifstein (2017).
86 Neoliberalism, as explained by geographer David Harvey, proposes that "human well-being can best be advanced by liberating individual entrepreneurial freedoms and skills within an institutional framework characterized by strong private property rights, free markets, and free trade." Neoliberalism is a theory of political economic practices that "seeks to bring all human action into the domain of the market" (Harvey, 2005: 2–3).
87 Watts (2004: 199).
88 Louisiana State Legislature (1978).
89 Moskowitz (2014).
90 Williams (1979a: 200).
91 Conversation with Chairman Chuckie Verdin in Montegut, Louisiana, April 4, 2012.
92 Austin (2006: 677).
93 Conversation with Victor in Isle de Jean Charles, Louisiana, July 14, 2012.
94 This concept is based on Marx's notion of primitive accumulation, which entailed "divorcing the producer from the means of production" (1994/1888: 296).
95 Story circle, Isle de Jean Charles, Louisiana, March 15, 2012.
96 Austin (2006).
97 Briscoe Center for American History (2014).
98 BP (2014).
99 Juhasz (2011: 213).
100 Westerman (2002).
101 Ostrom et al. (1999).
102 Peterson (2011).
103 McNeil (2011: 121).
104 Harvey (2005).
105 The Apache Corporation's name does not have any connection to the Apache tribe. The corporation's name comes from the founders' initials with "che" added at the end (Apache Corporation, 2014a).
106 Apache Corporation (2010: 4).
107 Apache Corporation (2014b).
108 Conversation with Chris Brunet in Isle de Jean Charles, Louisiana, April 11, 2012.

5

COREXIT TO FORGET IT

The BP Deepwater Horizon Oil Disaster

> My people were here first, and now the White man and government are
> damaging the place we call home with rules that take away what is ours for
> their own benefit. The oil companies have ruined our lands with their greed
> and lack of regulations. . . . BP was not the first oil spill just our biggest. BP
> is telling everyone that the people from the bayous are fine. They are lying.
> It has taken away our livelihood. Our seafood that feeds our families is no
> longer safe.
>
> —Crystlyn Rodrigue, Grand Caillou/Dulac[1]

This chapter highlights the 2010 BP Deepwater Horizon Oil Disaster—the
most recent infamous event in a centuries-long story of smaller-scale environmental and social disruptions. More critically and specifically, the chapter
examines the use of Corexit dispersants in an attempt to ameliorate contamination that further polluted the environment and affected local residents' health
and livelihoods.

★ ★ ★

Sunrise

By Crystlyn Rodrigue, Grand Caillou/Dulac[2]

In the summertime, the early morning is usually the time my dad and I set out to
go shrimping. When I first walk out of my house the air frequently catches me
off guard. The hit of the cool air shocks me because it is summertime. If you ever

lived in South Louisiana you should already know how scorching hot it is during the summertime. So, to get ambushed with some cool air is always nice. The sky is usually still black with stars practically covering the sky. The moon is big, full, and bright, brighter than the streetlights across the street from my house all along the highway. My dad and I always walk down our steps across our yard to the boat together with the moonlight guiding our way.

One at a time, we climb into the boat. My dad sits in the back of the boat behind the wheel, and I sit in the front in the picking box. Halfway asleep, I wrap myself in a blanket. Depending on how fast we ride the wind can get pretty cold. The boat ride starts off slow. The sound of the motor is a loudness that can probably bust an eardrum. It may sound like a painful noise, but for me it is a strangely calming sound.

I've always observed my surroundings. From the bayou, I look at the back of everyone's houses. On each side of the bayou no matter what the house looks like they have either one or two things—a wharf or a boat. During the ride, we go under three bridges before getting to the lake. The closer we get to the lake, the faster we get to ride. The faster we ride, the colder the wind. We ride so fast that it looks as if we are not moving at all. There is nothing sweeter than the smell of the open lake. Refreshing, yet salty. When we get to where we want to go we come to a complete halt.

My dad, behind the wheel, takes a look at the open water to find the "spot." The spot where the shrimp are. The look and the intensity on his face never changes. His eyebrows are together with a spotlight in his hands, which always means he is serious. After a long hard look I always ask him the same question, "Where are we going?" Meaning, where are we dropping the nets. My dad knows the water like the back of his hand. No matter how much the land and waters have changed over the years he still knows exactly where he is. It amazes me how smart he his being that he only has a fifth grade education. He is the smartest man I know.

We pick a spot to settle down. The nets are in the water. The sky starts turning a light blue. The stars start to fade. The moon gets higher in the sky. This is the moment I wait for. Smelling the crisp air. The boat slowly moving. The shrimp jumping to their own beat. I sit down next to my dad while he watches the wheel. I cherish the moments I have with my father, and shrimping has always been a big part of those moments.

I've always seen color in the world, but now that I am of age the world isn't as colorful as I thought it was. I'm 19 and my colorful world is slipping away, along with our way of life. My people were here first, and now the White man, and government are damaging the place we call home with rules that take away what is ours for their own benefit. The oil companies have ruined our lands with their greed and lack of regulations. This has been going on for many years, and I am just now seeing it.

BP was not the first oil spill, just our biggest. BP is telling everyone that the people from the bayous are fine. They are lying. It has taken away our livelihood. Our seafood that feeds our families is no longer safe. When President Obama came here he only went to one affected community hours from my home. He didn't come here or all of the other affected bayous. He did not say what will happen when the next hurricane comes and brings all that oil into our homes. What are we supposed to do then?

With each storm that passes we never know what we will come back to. It has become our norm to just come back, clean up, and start all over again and again and again. I recently asked my mom, "Would it be easier to just relocate?" Some days I do think it would be easier, but then I really think about it and ask myself, "Would it really be easier?" This is my so-called life.

My way of life is here, my people are here, this is who I am. I just want be on the boat with my dad shrimping. We have our most extraordinary conversations. There is advice when I am feeling insecure. Storytelling about when he was growing up here, about my grandfather Papa Tig, about them shrimping together. Indoors he is a quiet man, but when we are outdoors on the boat together I am filled with his words of wisdom. As the time passes on the sun starts to peak over the trees in layers of yellow, pink, and orange. Finally, the sun rises. Our day has only just begun.

FIGURE 5.1 Crystlyn Rodrigue's house flooded during 2011 Tropical Storm Lee, Grand Caillou/Dulac.

Source: Photo courtesy of Chief Shirell Parfait-Dardar, 2011.

FIGURE 5.2 Graffiti art, in response to the aftermath of the Deepwater Horizon oil rig explosion, Bayou Lafourche.

Source: Photo by author, 2010.

★ ★ ★

The BP Deepwater Horizon Oil Disaster

On April 20, 2010, eleven men were killed when BP's Deepwater Horizon oil rig exploded. Nearly five million barrels of oil spilled into the Gulf of Mexico before the well was capped on July 15, 2010, affecting approximately 1,100 miles of coastal wetlands.[3] BP initially estimated that between 1,063 to 14,226 barrels of oil spilled into the Gulf per day.[4] The 1972 Clean Water Act applies penalties for each spilled barrel; in addition to protecting its public image, it financially behooved BP to underestimate the spill's size.

The Response

Following the Deepwater Horizon Oil rig explosion, Louisiana Governor Bobby Jindal pushed the US Army Corps of Engineers (USACE) to approve a sand berms project that was supposed to prevent oil from reaching the marshes. Despite commenting agencies and scientists expressing concern that the berms would not be constructed in time to be effective and could potentially cause even further environmental damage, the USACE approved a scaled-back 39.5-mile berm project. The USACE estimated the cost of the project at $424 million. However, according to the National Commission on the BP Deepwater Horizon Oil Spill and Offshore Drilling (2011: 157), only "a fraction of the planned reaches would be finished before the spill ended, and very little oil would be captured."

Part of the problem with the haphazard cleanup was BP's Oil Spill Response Plan for the Gulf of Mexico. Portions of the response plan were copied from material on the US National Oceanic and Atmospheric Administration's (NOAA) website and failed to consider the applicability of the information it contained to the Gulf of Mexico context. In the Plan, BP named a wildlife expert on whom it would rely for expertise, yet this person had passed away several years before the Plan was submitted.[5] The US Minerals Management Service approved the error-filled Plan. The approval of oil spill response plans without close attention to important details is systemic throughout the oil and gas industry; ExxonMobil, Chevron, Cono-coPhillips, Shell, and others drilling in the Gulf of Mexico submitted nearly identical response plans prepared by the same contractor, all approved by the Minerals Management Service.[6]

As is often the case across extractive industries, regulators and agency and corporate representatives turn a blind eye to the potentially disastrous risks to workers and local communities and gross rate of infractions in exchange for economic gain. Between mid-2007 and early 2010, BP accounted for 862 safety citations from the Occupational Safety and Health Administration, which was nearly half of all citations to the entire refining industry.[7] The federal government had failed to guarantee agency regulators the political autonomy to enforce and overcome the oil and gas corporate interest that continued to oppose stricter safety regulations.

In 2014, a US District Court of Eastern Louisiana judge issued an historic ruling stating that BP exhibited "gross negligence" and "willful misconduct" leading up to the BP Disaster. However, with governments continuing to permit oil and gas corporations to explore and drill in increasingly precarious environments (e.g., the Arctic), hydraulic fracture for natural gas around sacred sites (e.g., Chaco Canyon), extract toxic unconventional hydrocarbons (e.g., Alberta tar sands), and violate the human rights of local communities (e.g., the Niger Delta), it is yet to be seen if the 2014 ruling of "gross negligence" indicates any real and lasting change.

Disaster Capitalism

The sociohistorical and political processes that established an energy sacrifice zone and privatized common resources perpetuate through the tenets of disaster capitalism, which have enabled corporations' ability to not only reap rewards, but to do so expeditiously. Disaster capitalism, for example, was seen in the United States through the "shock waves" sent through New Orleans immediately following Hurricane Katrina, or, rather, the "disaster *after* the event that reproduced social inequalities, in large part through the process of disaster capitalism."[8] Two different communities emerged in New Orleans post-Katrina. One community was the Bechtel or Fluor built trailer camps for low-income evacuees, administered and patrolled by private security companies.[9] By contrast, the other type of community emerged in the economically wealthier areas, where residents had water and emergency generators within weeks after Katrina hit, people were treated in private hospitals, and children attended new charter schools.[10] Political scientist Adolph Reed Jr. (2008: 148) explained, "The people who were swept aside or simply overlooked in this catastrophe were the same ones who were already swept aside in a model of urban revitalization that . . . is predicated on their removal."

Disaster capitalism can be defined as "national and transnational governmental institutions' instrumental use of catastrophe (both so-called 'natural' and human-mediated disasters, including post-conflict situations) to promote and empower a range of private, neoliberal capitalist interests."[11] Through such empowerment, opportunities for radical policy reform are created that constitute what humanitarianism researcher Antonio Donini (2008) called "world ordering." This type of "world ordering" and resource control is illustrated in coastal Louisiana through the habitual disasters and land loss the bayou communities experience across time, and most readily through BP's and the government agencies' response to the BP Deepwater Horizon Oil Disaster.

Corexit to Forget It

Substantiating my argument that government/industry involvement has created a sacrifice zone, the US Environmental Protection Agency (EPA) approved using dispersants—Corexit 9527A and then Corexit 9500A—below the surface of the water for the first time following the Deepwater Horizon explosion,[12] applying approximately 1.84 million gallons of dispersant to the Gulf waters by boats and airplanes.[13] Dispersants do not entirely remove oil from the water, but rather work in conjunction with the wind and waves to accelerate dispersal by allowing the oil to mix with water. The cleanup approach could be seen as matching the way local communities are at times made invisible by government/industry—out of sight, out of mind. The use of dispersants, Nicholas, a life-long fisherman from Pointe-au-Chien, remarked during a conversation on his porch about two years after the offical start of the BP Disaster, made a 4-by-6-mile-wide spill sink and disappear.[14]

BP chose to spray dispersants across the water's surface and inject directly at the well site. They chose the chemical cocktails of Corexit to sink the oil, which they were empowered to do through the Toxic Substances Control Act; the Act instituted that companies do not have to prove the safety of substances they release into the air or water, or even, in most cases, disclose what is in the products.[15]

Subsequent to the decision to use Corexit, EPA studies found that of the 18 approved dispersants, 12 are more effective on southern Louisiana crude oil than Corexit, and the toxicity of those 12 was either comparable to Corexit or, in some cases, even 10 to 20 times less toxic.[16] Corexit was on the EPA's National Contingency Plan Product Schedule, but the testing it had undergone did not consider the product's long-term impacts.

Corexit 9527A and 9500A contain propylene glycol, and Corexit 9527A contains 2-butoxyethanol (2-BE). Both are toxic, and both bioaccumulate up the food chain.[17] In fact, 2-butoxyethanol was identified as a cause of chronic health problems and even several deaths among cleanup workers after the 1989 Exxon Valdez

Spill in Prince William Sound, Alaska.[18] The health effects on Gulf residents have also been palpable. Victor (Isle de Jean Charles) recalled,

> We was working out there when they sprayed that dispersant. . . . My memory since that oil spill, that stuff I tested positive, I got a loss of memory . . . they're gonna pay you for your health . . . but the money's not gonna do much good. . . . I never had much money in my life and what's it gonna do now? I'll be sick and I won't be able to enjoy it when I'll be going down the road and forget what I'm doing.[19]

The full potential health effects of Corexit are unknown because the dispersant's manufacturer has refused to reveal all of its ingredients, citing their proprietary nature.[20] A University of South Florida study found that Corexit broke the oil droplets into smaller drops and created a plume that caused the die-off of foraminifera—amoeba-like creatures that are the basis of the Gulf 's aquatic food chain.[21] One recent study found that adding Corexit 9500A to the oil spill in the Gulf made the mixture up to 52 times more toxic than the oil itself.[22]

Nalco Company, which manufactures and sells Corexit, and includes a BP board member among its executives,[23] included in its portfolio "technologies that increase production, reduce operational costs and protect assets in challenging environments like Deepwater & Ultra-Deepwater, Oil Sands, and High Temperature High Pressure Corrosion. We also have chemistries designed to treat the heaviest crudes and oil spills."[24] Nalco thus supports drilling under more precarious circumstances, while also selling products to clean up the spills caused by such drilling.

Further, less than two weeks before the Deepwater Horizon oil rig exploded, Halliburton—the contractor responsible for cementing the rig's well—made a deal to purchase the firm Boots and Coots, which focuses on oil spill prevention and response to blowouts. Following the Deepwater Horizon blowout, this same firm was hired, under contract to BP, to assist with the Deepwater Horizon relief well work.[25]

The Claims Process

The 1990 Oil Pollution Act deemed that a private company responsible for a spill would be required to plug the well, clean up any resulting pollution, and compensate the people affected. BP established a $20 billion fund to compensate individuals and businesses for environmental and economic damages, subsistence losses, and property damage, as well as state and local agencies for response costs.[26]

I spoke with the consultant assisting Isle de Jean Charles and Grand Caillou/Dulac with their claims process. He noted that when he first met with Kenneth Feinberg, the BP and government-appointed fund administrator of the BP Deepwater Horizon Disaster Victim Compensation Fund, Feinberg made the process sound so simple and easy. But it had been a complete nightmare.[27] The lack of transparency in the claims process, denial and inadequacy of compensation, and

Feinberg's perceived influence from BP left people affected by the disaster frustrated and confused, furthering the damage caused.[28]

Feinberg rejected approximately two-thirds of the 480,000 claims he received (Hammer, 2011). BP mandated that local fishermen—who lost their livelihood because of the oil spill—hired for cleanup activities had to sign an agreement that affected their future potential legal claims.[29] Senator Landrieu (2011) from Louisiana said at a Congressional hearing on the claims process, "It is clear to me . . . that the law is deficient and can be and hopefully will be corrected so that the next time there is an environmental spill of a significant magnitude where there are impacts, not just environmental, not just economic, but community impacts or human service impacts, that the polluter, the violator in this case be held accountable."

In 2011, BP spent $8.43 million lobbying the US Congress, including working to block a bill that would have raised the liability cap oil corporations are responsible for in the event they are accountable for a major disaster.[30] During the first quarter of 2011 alone, BP spent $2 million on federal lobbying for issues such as capping its contributions to the restoration of the Gulf Coast.[31]

Nearly three years after the spill, in December 2012, I watched Isle de Jean Charles' Tribal members fill out claims forms, which asked for paperwork and past fishing licenses that most people no longer had. To be eligible for compensation, BP required three prior years of paystubs, paychecks, tax returns, and documents created prior to the 2010 Disaster. However, many people had lost these items due to flooding during Hurricanes Gustav and Ike in 2008, only two years before the BP Disaster. On top of this, many activities related to subsistence claims were conducted informally and had no written record. Local people were frustrated that no one seemed to know the status of the claims process. With layers of disasters and events occurring in the region, it was exceptionally challenging to document a direct link between subsistence and livelihood impacts and the BP Disaster,[32] especially when effects could occur in any number of places along the aquatic food chain.

Some residents talked about tension created in communities around the claims process and who was hired for the cleanup work. For example, Patrick, who had relocated about 15 miles northwest from Pointe-au-Chien to Montegut, did not want to take part in the claims process because he felt it was causing unnecessary conflict in the community: "That's what they do, they come in and tear people apart so it's harder for them to band together."[33] However, some people felt that communities pulled together during and after the cleanup.

The Forgotten Bayous

I visited the Louisiana Bayou region in June 2010, less than two months after the Deepwater Horizon oil rig exploded. Chief Albert took me out on a boat around Isle de Jean Charles and showed me the boom—a floating barrier—that cleanup workers were placing around the Island, to try to prevent the oil from reaching the Island (Figure 5.3).

During that same visit, Theresa (Pointe-au-Chien) explained that the local parish officials did not visit her community following the spill. BP workers came to the

FIGURE 5.3   Cleanup workers place boom around Isle de Jean Charles to try to prevent oil from the BP Deepwater Horizon Oil Disaster coming ashore.

Source: Photo by author, 2010.

area only because the Tribal leaders brought them there to set up an incident command center to try to prevent the oil from reaching the Tribes' lands. She expressed that they were the "forgotten bayous."

Starting my full-time research on the ground one and a half years after the BP Deepwater Horizon Disaster officially started, it was less about what I saw and more about what was missing—shrimpers *not* shrimping, crabbers *not* crabbing, people who relied on seafood to sustain themselves *not* eating seafood. Lifelong fishers had to park or sell their boats, losing a way to provide a meal for their families and community, and signaling a loss of income and way of life.

Shortly after the second anniversary of the official start of the BP Disaster when the Deepwater Horizon oil rig exploded, I sat with Nicholas (Pointe-au-Chien) on his elevated porch looking across the narrow road at Bayou Pointe-au-Chien. He talked about how people were affected by the lack of catch since the oil spill:

> More anger now. They get riled up quick. You got erosion. It might get worse before it gets better. But hopefully the seafood picks up, put a little smile on some people's faces. It's like a therapy session when you go out. It does something to you. When you ain't catching nothing it gets people more angry. It's not no good therapy no more. You goin' out there it's like going to the casino. It's a gamble and you hope you gonna win.[34]

As we talked, he gazed out across the bayou at the dock where his boat sat, where his bathtub that was typically filled with crabs shedding their shells rested. Nicholas told me this was the first year he was not crabbing because the crabs were too small, yet he still had to pay Apache Corporation and ConocoPhillips to lease a section of water to put his crab traps out.

He pointed to his dock and described how life had changed since the BP Disaster: "Used to crab boil there. Every two to three days was boiling crab. That's a big difference for me. I mostly now sit alone." He no longer took his seafood to market. Each time the phone rang with buyers, he took down their information but told them he had nothing to sell. During our conversation, he kept coming back to hoping things would get better. But, as he said, his therapy of being out on the water fishing was gone.

He talked about a boat blessing that took place a few weeks before our conversation, when family and friends gathered together to go down the bayou on boats as a priest blessed the boats for the start of the shrimp season. This year marked the first time they did not have a lot of crabs and there were no shrimp for the celebration. I recalled seeing the giant cooler on his brother's boat filled with crawfish and another half-filled with crabs. But I am not from here. I did not know to look for the shrimp. I did not know the crab cooler should be full.

The disaster accelerated the trend of an already declining shrimping industry due in large part to industrial restructuring and aquaculture imports of farm-raised shrimp from low-wage producers in other countries.[35] Going out into the waters had become too financially risky for local shrimpers because their catch was way down, the cost of fuel and ice had drastically increased in recent years, and the price shrimpers received at the dock had substantially dropped because of the influx of imported, low-priced shrimp.[36] Due largely to the large farmed shrimp boom in Southeast Asia and Latin America, world prices for shrimp dropped from $6.79 per pound in 2000 to $3.49 in 2009.[37] When shrimpers in Louisiana sold at the dock, they received less than $2 per pound.[38] Further, inshore commercial operators, who rely on ice to keep their shrimp chilled, reported spending $500–$800 per trip on ice.[39]

When I was out on the water with local shrimpers a couple years after the spill, their disappointment was palpable as they compared the thousands-of-pound catches of the past to the now typical couple-hundred-pound catch that barely allowed them to break even. A number of local fishers expressed no longer finding the same joy they once did in fishing, as they never knew what they would bring back, if anything at all.

Several people described to me all the places around the country they had family members they provided with seafood. Their inability to stockpile and send food to family members who had moved elsewhere resulted in those living outside the region losing access to local seafood supplies and the those-cultural connection of sharing and resource exchange, and connection to place and community.

The Disaster Narrative

In early 2010, when the Deepwater Horizon oil rig exploded, BP was the largest oil and gas producer in the Gulf of Mexico and the United States as well as the fourth-largest corporation in the world by revenue.[40] BP made almost $16.6 billion in profits and $239 billion in revenues in 2009,[41] the year before the BP Disaster officially began. Between April and July 2010, BP spent nearly $93.5 million in advertising, more than three times the amount spent during the same period in 2009. Some coastal residents and fishing families took action to counter BP's media campaign, attempting to influence what anthropologist Mark Schuller (2016) described as the "disaster narrative." For example, Nicholas noted how he drove around in his boat taking pictures "in case they tried to say it didn't happen over here. Know they'd try to get out of paying people, so I have pictures." Others spoke out publicly, refuting the assertions made by the media campaign—namely that no, the disaster was not over. While locals could not always pinpoint exactly what was happening, they did know that what they were seeing and experiencing was different from anything they had seen before.

Audrey, who lived in Montegut, about 15 miles north of her family in Pointe-au-Chien, told me one day while we sat at her kitchen table,

> I got some shrimp from one of my cousins, I think it was around the begin-
> ning of the shrimping season . . . and what I found interesting with the
> shrimp is the water they were in was a lot darker, it was almost black, and
> the heads, they had almost an oily feel to them, the substance . . . I had to
> wash and wash and wash them to get them clean, and keep washing them
> until your water runs clear, but I don't remember shrimp in the past that I've
> cleaned before being that dark, so I don't know.[42]

She noted how her family members talked about not seeing eggs in the shrimp like they should, skinny oysters, and no crabs. "What's caused it? Who's to say? Is it the dispersant? Is it a combination of the dispersant and the oil?"

A few months later, at a July 2012 Gulf Organized Fisheries in Solidarity and Hope Coalition (GO FISH) forum in Houma, a local fisherman voiced concern about recently seeing shrimp caught without eyeballs. Like a wave around the room, others murmured, "We have some too" and "We all have some." An elderly shrimper observed that the 2012 season was the worst catch he had ever seen. Another resident noted that before the spill he would always see oyster shells filled with baby oysters, but now the shells were turning black and once that happened the shell was dead. A young crabber stood up at the community forum to say that the compensation he was receiving was "like a slap in the face . . . our livelihood is priceless." The greatest uncertainty was not knowing what they would see the following year as impacts on fish reproduction, development, and populations might take years to determine. For instance, new developments from the 1989 Exxon Valdez Spill in Alaska show the long-term effects on herring and salmon from that spill are proving to be worse than originally acknowledged.[43]

The BP Disaster raised public concerns about the safety of seafood from the Gulf of Mexico. Government officials made statements confirming the safety of seafood harvested from the portion of the Gulf that had been reopened after being closed immediately following the spill.[44] However, what the government and BP were saying about the seafood was distinctly different from what local fishers and residents

FIGURE 5.4 Oyster shells covered in oil near Pointe-au-Chien.

Source: Photo courtesy of Babs Bagwell, 2012.

were finding. Locally observed deformities in seafood species matched research studies documenting fish with lesions and deformed shrimp.[45] Human health concerns were vindicated when high levels of compounds present in crude oil released during the BP Disaster were found in shrimp and oysters in the northern Gulf of Mexico (Figure 5.4).[46] Crude oil contains high levels of volatile organic compounds like benzene, a known human carcinogen and the cause of other serious health effects.[47]

When the government tested the seafood in the area it used studies based on the national average consumption of seafood, not for people who eat it on a daily basis for the majority of their dietary intake. It can be argued that the US Food and Drug Administration underestimated the risk from seafood contaminants, including known carcinogens and developmental toxins.[48] Residents were no longer sure of what they were putting into their bodies.

Toxic Uncertainty

The BP Deepwater Horizon Oil Disaster exacerbated the "toxic uncertainty"[49] that residents were already experiencing, which stemmed from multiple sources of contamination, unknown toxic substances, and the confusion caused by multiple discourses and actors intervening. Community members talked to me about changes in their own and family members' health over the years, including soaring rates of diabetes, cancer, and high blood pressure. Decades-long industrial contamination, encroaching toxic industries, chemicals from dispersants, oil spills, and post-storm debris continue to contaminate the communities' air, soil, and water, creating severe health and livelihood effects, along with a forced change in diet.[50] There are cancer treatment centers and hospitals in nearby cities, but this does nothing to resolve the cause of the cancers in the first place. Feeling a distinct loss of power, residents often avoided thinking about these issues unless they were provoked by an event, such as an oil spill or a researcher asking questions.

Natural gas, oil (hydraulic, diesel, crude), drilling mud, hydrogen sulfide and cyanide, ethyl compounds, and sulfur dioxide from the oil and gas industry cause a slew of severe health issues, including developmental, respiratory, digestive, neurotoxin, renal, and dermatological.[51] The Louisiana Bucket Brigade—a non-profit environmental health and justice organization—has been conducting long-term monitoring of air quality in the region and has also charted an increase of reported spills and gas released in the region. In 2009 there were 3,636 reported spills/releases; that number increased to 4,888 in 2010. An estimated 214,000 pounds of air pollutants are generated by every offshore oil platform each year,[52] polluting the air around the tribal communities. All these releases have serious impacts on human and environmental health.

Kerry St. Pé, the Executive Director of BTNEP, described his findings working on water pollution control for Louisiana Department of Wildlife and Fisheries.[53] He found that the oil industry was discharging 89 million gallons of produced water (a waste byproduct of the oil and gas industry) per day along the coast of Louisiana, and the average level of radium-226 in the produced water was orders of magnitude higher than the allotted discharge from nuclear power plants. It sank not just to the bottom but also into the bottom, into the same water and ecosystem that feeds and sustains the people along the coast. One of the naturally occurring radioactive materials in produced water from oil and gas production is radium-226, which can cause cancer in people if inhaled or ingested.[54]

Not only do many local residents fear the contamination in their drinking water, but in their soil as well, in part because of the heavy metals that come with flooding and the encroaching oil industry. Previously, longevity was a part of their heritage, but now, as was expressed at the January 2012 workshop in Pointe-aux-Chenes about what climate, weather, and other environmental changes meant for the communities, "we are the sacrificed communities and our people are dying younger because of new diseases we never had before."[55]

Tar balls still washing ashore and into the marshes five years after the spill contained *Vibrio vulnificus*, a human pathogen, which can cause severe wound infections, with a 20–30 percent fatality rate of patients contracting infections due to exposure.[56] With chemicals from dispersants still infusing the Gulf's waters, fishers are experiencing bacterial infections years after the spill, but finding it increasingly difficult to attribute them to a particular source.

Besides the health effects already being experienced, a number of people were concerned about potential health issues to come. For example, when I visited in

June 2010, Donald, the Co-Chairman of Pointe-au-Chien, took his boat out for the cleanup, but was careful not to touch the oil. When he later power-washed his boat, the copper paint came off, which normally only happened when it was sanded off. Some scientific experts on oil spills were worried that boats used during the cleanup could be soaked with chemicals that might seep through the boat's wood and harm the people working on the boat.[57]

Four years later, standing on the deck of a docked shrimp boat at the southern end of Pointe-au-Chien in late 2014—four and a half years after the spill—a local shrimper friend told me why he had not shrimped much that season. A couple months prior he was cleaning his boat with the same chemicals he had used for years, but something in the water had mixed with the chemicals, went through his boot, and made him sick for weeks. He told me another fisherman friend had recently passed away due to a similar incident.

While some of the health issues were not life threatening, they were new and chronic. When I asked residents if they had experienced health problems since the 2010 BP Disaster, often they would say no, but then a few minutes later say how they had been having chronic issues that arose in the previous two years, such as sinus and breathing problems. Decades of living with contamination seemed to have normalized such issues and made them part of everyday life.

In June 2012, Crystlyn (Grand Caillou/Dulac) and I stood barefoot in the water across from the small, eroding beach in Dulac where her family and others used to gather after the springtime boat blessing. We tried to avoid the occasional crab nipping at our toes. She voiced concerns about what would happen if another hurricane hit and brought in the oil. She said there was a reality here people did not want to face so they did not talk about it. Tears welled up in her strikingly dark, youthful eyes and she looked straight into mine. I started to say it was going to be okay. But I couldn't.

Less than three months later, in late August, I stood in the flooded water in front of her family's elevated trailer a couple of days after Hurricane Isaac hit. The same fears resurfaced. Later that same day, navigating through the flood waters from Hurricane Isaac, I saw Chief Shirell on her way back up the bayou, returning from the southern end of Dulac to take pictures and put water in a plastic bottle to be tested for oil and Corexit dispersant. She saw bubbles when the water was going back out into the bayou that she had never seen before. She did not know what it was, just that it was different. The uncertainty local residents had been expressing since the spill was pervasive: what happens if a hurricane comes and stirs up the oil and chemicals that were sunk?

The words that Chris (Isle de Jean Charles) shared with me months before resonated more than ever: "If a storm brought oil into our homes and the government said we couldn't go back, I couldn't put the impact into words. I still want to live here because I'm Native American, because I'm connected to the land. I'm going to live here as long as I can. I belong here."

The following chapter details the ways the Tribes both resist the adverse events forced upon them, and adapt to the environmental changes and the layers of disasters.

Acknowledgment

A special note of gratitude to Crystlyn Rodrigue for contributing her story "Sunrise" to this book and for our conversations and moments spent together.

Notes

1 "Sunrise," story by Crystlyn Rodrigue, Grand Caillou/Dulac, August 14, 2012.
2 Story, August 14, 2012.
3 National Commission (2011); NAS (2013).
4 Ibid.
5 National Commission (2011).
6 Ibid.
7 Freudenburg and Gramling (2011: 42).
8 Schuller and Maldonado (2016) (italics original).
9 Klein (2007: 524).
10 Ibid.: 532.
11 Schuller and Maldonado (2016).
12 Center for Biological Diversity (2014).
13 United States Coast Guard (2011).
14 Conversation with Nicholas in Pointe-au-Chien, Louisiana, May 4, 2012.
15 Sheppard (2010).
16 Quinlan (2010).
17 Subra (2010); Center for Biological Diversity (2014).
18 Juhasz (2011: 100); Center for Biological Diversity (2014).
19 Conversation with Victor in Isle de Jean Charles, Louisiana, July 14, 2012.
20 Center for Biological Diversity (2014).
21 Pittman (2013).
22 Rico-Martínez et al. (2013).
23 Quinlan (2010).
24 Nalco (2014).
25 Rocheleau (2010); Schuller and Maldonado (2016).
26 Landrieu (2011); Weber (2011).
27 Conversation with a BP claims consultant in Pointe-aux-Chenes, December 15, 2012.
28 Associated Press (2011); Weber (2011).
29 Subra (2010).
30 Stein (2012).
31 Baram (2011).
32 Gulf Coast Claims Facility (2011).
33 Conversation with Patrick in Montegut, Louisiana, June 6, 2012.
34 Conversation with Nicholas in Pointe-au-Chien, Louisiana, May 4, 2012.
35 Harrison (2012).
36 Ingles and McIlvaine-Newsad (2007).
37 Mine et al. (2016).
38 LDWF (2016).
39 Mine et al. (2016).
40 Juhasz (2011: 213).
41 Ibid.
42 Conversation with Audrey in Montegut, Louisiana, January 14, 2012.
43 Steiner (2015).
44 NOAA (2010).
45 Jamail (2012).
46 Sammarco et al. (2013).

47 Juhasz (2011: 90); Solomon and Janssen (2010: 1118).
48 Rotkin-Ellman et al. (2012).
49 Auyero and Swistun (2009: 66).
50 Coastal Louisiana Tribal Communities (2012); Maldonado et al. (2013).
51 Lasley (2011).
52 Juhasz (2011: 257–258).
53 Conversation with Kerry St. Pé in Thibodaux, Louisiana, July 27, 2012.
54 EPA (2012).
55 Coastal Louisiana Tribal Communities (2012).
56 Strom and Paranjpye (2000: 179); Tao et al. (2011).
57 Juhasz (2011: 190).

6

ADAPTATION AND RESISTANCE

Our people have always lived off the water and land. We're bayou people. After a storm, sometimes there's still water on the road, but we come back. People here come back. It's like nowhere else.

—Theresa Dardar, Pointe-au-Chien[1]

Despite the layers of disasters and accumulating effects, the Tribes continue to adapt to the changing environmental conditions, both in individual and collective ways. Much like the continued flow of water up and down the bayous, the Tribes' adaptation is often subtle, quiet, but persistent. Not all residents pursue the same adaptation strategies or have the same vision of adaptation. However, many residents undertake different forms of everyday adaptation, whether it is putting oyster shells around their house or continuing to come back and clean up after each hurricane. People's adaptation efforts can be both self-indulgent and revolutionary, with the intent for survival, as well as immediate- and long-term gains. In addition to the environmental changes, many residents both resist against the dominant sociopolitical and economic structures but also find ways to adapt and survive within them.

This chapter details some of the resistance actions tribal members pursue to advocate for their communities and the strategies they practice to adapt to changing environmental conditions.

★ ★ ★

I Came and I Stayed

By Theresa Dardar, Pointe-au-Chien[2]

I was born in New Orleans and raised in Houma, Louisiana in a segregated community with Whites at the beginning of the street, Indians in the middle and Blacks at the end. One day after seeing a little girl get whipped for coming to play with us, I asked my grandfather why her mom whipped her. My grandfather told me that most White people did not like Indians. He also told me never to be ashamed of who I am, but to be proud and never deny it. He said that God sees all and like God we should not see colors.

I went to an all-Indian school growing up. I still remember my first grade teacher because she's the one who taught me English. In seventh grade, after desegregation happened I became a different person because the Whites would call me a *sabine*, a derogatory word for Indian. I would end up in a fight and get suspended. I failed my first year at Oaklawn Jr. High and quit in ninth grade. At sixteen, I went to work at Val-U Department Store and worked there until after Donald and I were living together.

Moving 20 miles southeast from Houma to Pointe-au-Chien was a very big difference. I had a longer drive to work and had to relearn French to talk with Donald's family. Although I was raised speaking French, I had forgotten much of the language because we could not speak French at school without being punished. I guess because the teachers were all White.

After living together for a few months I quit my job and started to go trawling with Donald. I liked being out on the water, doing something different than what I was used to doing. When shrimp season was over, Donald, his brothers and their dad would start trapping. Donald's dad skinned the muskrats, his mom skinned the nutrias and raccoons and put them on a board to dry. I didn't know anything about trapping or working the animals they caught, so I took care of Donald's youngest siblings and did house work. Then they decided it was time I learn to put the skins on the board. It was really a family affair because everyone did something. All the boys trapped and the girls pulled the guts out, cut the teeth, feet and tails off the nutria. They sold the meat for ten cents a pound and split the money.

It was a real change for a girl who grew up in the city never fixing fish, crabs or shrimp, not knowing anything about trapping, only hearing about it because my grandfather and uncle trapped. I was at least used to opening oysters because my dad was an oyster fisherman. I learned about trawling, driving the boat, helping Donald sort the shrimp and picking up the boards and the trawl. I once shook hands with a man and he told me that I had rough hands. I told him that was because these hands work. I can't work like I used to though because of my rheumatoid arthritis. Plus, there's no more trapping anymore. The nutria, muskrats and raccoons are not there and the price of fur has dropped.

This life is not an easy one, but I would not change my life because my roots are here in Pointe-au-Chien. My grandfather, dad and mom were born below the Cut Off Canal. Some of my ancestors are buried down the bayou and if it were permitted I would be buried down here instead of up the bayou.

A few years ago I became a Grail member, an international movement of women, and being with these strong women gave me courage to speak in public. I started doing Native American workshops and representing my community, the Pointe-au-Chien Indian tribe, at public forums, the United Nations, and by serving on many committees. I also work in our community and the Island community after hurricanes. I feel like I have done a lot of good for both communities because I know the people and know their needs. I try to be a voice for our small Native American fishing community.

The most important thing to me is trying to get the parishes to save what is left of our land, the land they want to see wash away so the oil companies can take over. When I was growing up, my family would take the boat down the bayou and I could pull the grass on either side of the boat with my hands. Now it is just wide open canals. It was pretty here before. Now all we have are skeletons.

Our community is made up of my people. We are all family. Everybody knows everybody. You can stop at the first house and ask where so and so lives. We don't lock our doors. We trust everyone here. Before the oil spill, you could go in front of your house and catch fish or crabs for supper or throw a cast net and catch shrimp.

You need your people and you need your land. But our land is slowly washing away. Without the land, our community will be separated. Our younger generation is leaving. Pretty soon we're going to be just an elderly community. The land, at least what's left, is what keeps our community together. If we scatter into other communities, we will lose our Indian bloodline. We want our children to be able to stay in the community to keep the Tribe going.

If the Morganza levee system isn't built further down we'll lose our cemeteries and ceremonial mounds that our ancestors built that are part of us, part of our people. The barrier islands need to be repaired and some of the canals closed that the oil companies opened up. Our ancestors went to the end of the bayous to save their lives so that they wouldn't be killed by the White man and now we suffer for it.

Our people have always lived off the water and land. We're bayou people. After a storm, sometimes there's still water on the road, but we come back. People here come back. It's like nowhere else.

★ ★ ★

FIGURE 6.1 Theresa Dardar, Pointe-au-Chien Indian Tribe.

Source: Photo by author, 2011.

FIGURE 6.2 Southern end of Bayou Pointe-au-Chien.

Theresa advocated for the Parish to put rocks in front of the tree on the right to protect the land, but no action was taken and the land washed away.

Source: Photo by author, 2012.

Recognition

One of the major challenges the Tribes face is the acknowledgment of their presence in the present. For example, at a meeting in New Orleans in March 2012, focused on new entrepreneurial ideas for Louisiana's "water challenge," I listened to US Senator Mary Landrieu discuss the importance of restoring the coast. She stated how Native Americans had been there many centuries before. Theresa (Pointe-au-Chien), sitting next to me, whispered that there was no recognition the Tribes were still here, referencing Senator Landrieu only speaking about the Tribes in the past tense.

In another instance, despite the US Census Bureau hiring people from the tribal communities to collect data for the 2010 Census, the 2010 American Indians and Alaska Natives US Census map labeled the entire tribal population in Terrebonne and Lafourche Parishes as "United Houma Nation," (UHN)[3] ignoring the fact that UHN is only one of five independent, state-recognized Tribes in the two parishes, including the three Tribes that are the focus here.

Residents advocate for acknowledgment in the present in multiple ways. The three Tribes are seeking federal recognition in part as a resistance against the erasure

FIGURE 6.3 Sign posted by an Isle de Jean Charles resident.

Source: Photo by author, 2010.

of their history (see Chapter 3). Some tribal members continue to speak out to save their land, as Chris (Isle de Jean Charles) described, to "keep us on the map," while others maintain different forms of demonstration, like putting up signs to announce that they are not going anywhere (Figure 6.3).

Tribal members attend public forums and meetings to show that there are still tribes living down the bayous and to advocate for their rights. For example, at a January 2012 public meeting in Houma for Louisiana's 50-Year Master Plan for a Sustainable Coast, which outlines the planned state-sponsored coastal restoration projects, Theresa stood up and said, "How do we know that our comments will be considered for real and not just to fit into your guidelines?" Standing there in front of the panel, in the middle of the crowded auditorium, she pushed back her long black hair, revealing the words written across her t-shirt, "Sure you can trust the government, just ask an Indian."

Reinvigoration of Cultural Traditions and Restoring Traditional Plants

The Tribes are reinvigorating some of their cultural traditions, such as forming a drumming group, holding naming ceremonies, building a traditional palmetto hut, and preserving their food and community traditions.[4] Cultural reinvigoration is exemplified during Pointe-au-Chien's week-long youth cultural camp during the summer with drumming, shawl making, basket weaving, drumming, beading, storytelling, and shared traditional food. During one camp, the children helped their elders and the Pointe-au-Chien Tribal Council and other Tribal members finish building a traditional palmetto hut, replicating the type of dwelling their ancestors lived in until the early 1900s.

Tribal leaders are pursuing planting traditional, medicinal plants and vegetables in raised-bed gardens to avoid the saline soil. They have planted more salt-resistant trees in the communities, with the support of a nearby US Department of Agriculture Plant Materials Center. Through partnerships with the US Department of Agriculture's Natural Resources Conservation Service (USDA-NRCS) and the Barataria Terrebonne National Estuary Program (BTNEP), some tribal members have recorded which plants have been lost and which ones still grow in the communities. This provides a better understanding of preservation pressures and identifies what can be grown in current soil conditions. However, this also means that the Tribes are forced into conducting a "cultural triage"—a forced choice to rank in importance cultural resources to be saved.[5] Less important resources could be saved as well by moving the resources elsewhere. For example, some of the medicinal and traditional plants could be grown in the regional USDA-NRCS Plant Materials Center. However, in growing the plants outside of the communities, the physical landscape for the cultural resources is broken.[6]

Pointe-au-Chien is working with an ethnobotanist to create a traditional medicine and food garden in the community and revitalize traditional, medicinal plant knowledge and use by making culturally important plants available and accessible to community members who will maintain the garden. The garden can serve as a focal point for gatherings on traditional plant use, such as the annual culture camp.[7] Some residents are also adapting individual planting practices in response to the newly saline soil. For example, Renée, an elder from Isle de Jean Charles told me, "I'm going to be planting as long as I live."[8] A few days later I watched her create a makeshift raised-bed garden out of an old cleaned-out toilet bowl to keep her plants above the soil.

Continuing their food traditions, to the extent possible after the BP Disaster, helps maintain their cultural identity and connection to place. Further, the tribal leaders are re-introducing traditions, establishing new rituals, and re-learning what has been lost. Michele, who had relocated from Pointe-au-Chien to Bourg, shared, "I don't remember my grandpa and them drumming. I'm sure like maybe in the 17, 1800s they drummed, I'm sure. But from what I remember from like my grandpa and them, nobody drummed. I find that that part we had lost."[9] Six tribal leaders from Isle de Jean Charles and Pointe-au-Chien formed a drum group a few years prior, having obtained the drum through a grant from the local Diocese. They practice together and perform at cultural events such as naming ceremonies and the Native American Mass held every year in Pointe-aux-Chenes.

Land Claims

Some tribal members who have relocated continue to claim the land on which they once lived. For example, in the annual Isle de Jean Charles Christmas Parade, as the parade goes down the Island and Santa passes out the donated gifts, many families stand where their houses used to be, now empty spaces grown over with weeds, remnants of what recent hurricanes left behind.

Reclaiming the land is about both present actions and memories of what their ancestors did to protect the land that sustained them and served as a safe haven. For example, Donald (Pointe-au-Chien, Co-Chairman) still holds on to a check his grandfather was given by outsiders for his land; his grandfather never cashed the check because he had not agreed to sell the land.[10] Some tribal members come together to lease some of the waters from oil corporations so their people can continue to access the water for fishing. They are also working to save and restore important places, such as their sacred mounds, despite facing obstacles with permitting and land ownership due to the state and oil companies taking over the land (see Chapter 4).

Protecting and Elevating Houses

Some people put oyster shells, rocks, and dirt around their houses for protection from erosion and flooding. Another strategy for adapting to increased flooding includes elevating the houses above the floodwaters. After Hurricane Juan hit in 1985, local residents started having their houses raised with the support of federal programs, religious organizations, non-governmental organizations, and private businesses. As flooding intensifies, the houses have to be raised increasingly higher. While this allows people to stay in place above the floodwaters, it is not without consequences. For example, the last house at the south end of Pointe-au-Chien belongs to a woman in her sixties. The house is elevated 19 feet, but does not have an elevator. Elevating houses higher puts residents, especially the elderly and those with special needs, at risk. People have injured themselves going down the steps, and are more vulnerable to the wind during storms. Yet, many people have been able to stay because they raised their houses.

Employing Local Knowledge

In coastal Louisiana, Native residents' intimate knowledge of the surrounding waterways and landscape enables them to see the changes happening and identify intervention strategies. As Nicholas (Pointe-au-Chien) noted, "I say we've been here all our lives. We know how the water works."[11] For example, Donald (Pointe-au-Chien, Co-Chairman) took a couple scientists out on his boat one afternoon to look into how the Tribe could restore its sacred mounds. I watched the scientists struggle to set up their instruments to measure the depth of the water. Donald kept saying the water was five feet deep. After several failed attempts while we sat there and waited amidst the swarm of mosquitoes, the scientists finally measured the water depth. It was five feet.

Knowledge Exchanges

Tribal leaders and community organizers work together, meeting with other communities and organizations to increase awareness and explore new options for restoration efforts. They combine their voices to be stronger advocates for their communities, interacting with and speaking at the United Nations, to federal agencies, and to numerous organizations and agencies along the Gulf Coast and elsewhere.

They have also created formal organizations. For example, through the USDA-NRCS, the three Tribes, along with Grand Bayou, another tribe in coastal Louisiana, created the First People's Conservation Council to address natural resource issues occurring on the tribal and coastal lands. The organization was established in 2012 through the tribal leaders and the support of the NRCS Wisconsin Tribal Advisory Council and the Lowlander Center, a local non-governmental organization with a long-standing relationship with the Tribes.

Tribal members have also organized learning exchanges with other tribes and communities, reaching out to those who are experiencing similar adversities. Participants share and learn from each other about environmental and resource challenges, adaptation processes, and problem-solving techniques. Through the support of the Lowlander Center and a nearby religious congregation, the Tribes hold fellowship gatherings with tribal representatives from Alaska and Minnesota, sharing knowledge and cultural traditions, such as basket weaving, drum-making, and ethnobotany, and experiences of extractive industries affecting their communities.[12]

The Tribes have connected with communities affected by the Exxon Valdez Spill in Alaska; tribal members have traveled to Cordova and Valdez, Alaska and vice versa to exchange stories. As a fellowship participant from Cordova reflected, "I came to the bayou with everybody and I looked at it and I went, oh this could be Cordova. . . . We have more similarities than we have differences. . . . It isn't just whatever person's story, it's the people's story."[13]

An important element of the knowledge exchanges has been the use of storytelling to build awareness, share knowledge, and increase solidarity across cultures, communities, regions, and even nations. Storytelling can lead to alliance building, through which inroads are made, resources are accessed, and capacities are built, contributing to a growing commons of information, expertise, strategic organizing, and problem-solving.[14] Building trust through collaborations and partnerships has helped facilitate the sharing of knowledge, lessons, and mutual aid.

Staying in Place

Despite the challenges, a number of people told me that they stayed in place because this was where their way of life was and they could not practice the same fishing traditions if they moved into the city. Many elders and adults had spent their entire lives working on boats, often reflecting they learned their way around a boat before they could even walk.

Residents of Pointe-au-Chien talked about how people always came back after storms because they liked the close-knit community, so they just kept rebuilding their houses higher. Theresa told me how one family in Pointe-au-Chien even slept on wet mattresses after their house had been condemned following a storm. When I asked Madeleine, an elder from Pointe-au-Chien who had relocated about 15 miles northwest to Montegut, what she thought about the Tribe's future down the bayou she commented, "Oh, the Tribe's not going anywhere. Some of them's going back. I've heard quite a few of them saying they're going back to Pointe-au-Chien."[15]

Despite the challenges of living on the Island, Pierre, an Isle de Jean Charles elder, echoed a similar sentiment when I asked him, over coffee on his porch, why it was important to stay on the Island: "I was born and raised over here that's why I live here. I've been all over the world and I came back here." Renée, who a few minutes prior had walked across the road and up the steps to Pierre and Marie's porch for afternoon coffee hour, noted, "I stay because I just love to be outside and not crowded. And look out. Well now I like to look out at the water. And watch the sunset and the sunrise. Can't see that in town. And you're all crowded. And everybody's on top of another. And that's why I'm going to stay as long as I can."[16]

While tribal residents adapt in both subtle and direct ways, they are still up against distinct challenges in terms of the rate of environmental change and the increasing accumulation of still recovering from one disaster when the next one hits, without sufficient governance support for proactive measures to reduce risk. A number of people often expressed that they would keep trying and not give up, but also had great frustration at the slow pace of restoration efforts unfolding in and around their communities, concerned that far too little was being done.

Restoration Frustration

State reports have concluded that without restoration and flood mitigation actions, much of the coastal tribes' lands would be gone before 2050, including all of Isle de Jean Charles.[17] Yet, the communities have thus far been mostly left out of state-led restoration, mitigation, and hurricane protection plans. The map for Louisiana's 50-Year Master Plan for a Sustainable Coast has a red line running through it to indicate where hurricane protection systems will be placed; the land north of the line will be protected from flooding and land loss by state-sponsored restoration and protection. Isle de Jean Charles is south of that line.

During a story circle at Pierre and Marie's house in Isle de Jean Charles in March 2012, I asked a few Isle de Jean Charles Tribal members who all grew up on the Island, but some of whom had relocated due to flooding, if they thought people would be able to stay on the Island or if the community would need to relocate:

Maurice: Well, I don't think they can save the Island.
Louis: They don't want to save the Island, they could if they wanted to.

Pierre: They could, but they don't want to . . .

Maurice: The levee's gonna pass south of Houma, you know where the Ranch Road is, the levee's gonna be right south of there. They started on it already.

Louis: So they're drowning everything below.

Maurice: South.

Louis: They're drowning everything south of that.

Pierre: South of Houma, yeah.

Louis: I would call that discrimination.

The snail-like pace of restoration adds to residents' frustration and sense of anxiety if enough will be done before the land in and around their communities is completely gone. For example, Chairman Chuckie (Pointe-au-Chien) told me about a newspaper article he read that discussed how the parishes could get money to fight erosion. However, he felt this was meaningless because the same thing had been said for years and the money was just used for studies: "Same thing for fifty years and nothing to show for it."[18]

Disempowered and cut-off from the restoration process, people's sense of dislocation and alienation increases as they continue to watch the lands around them disappear. Those who controlled the restoration process (e.g., government agencies) and those who own the means of production (e.g., oil and gas corporations, large-scale commercial fishing companies) often exclude the local residents from the process. As restoration becomes an object of commodification and racial tension, the tribal residents become further estranged and alienated from the physical environment and are denied the input of their local ecological knowledge, which could be an invaluable contribution to restoration.[19]

After releasing the 2012 Coastal Master Plan with the support of three focus groups—oil and gas, commercial seafood, and navigation industries—the Coastal Protection and Restoration Authority (CPRA), which is mandated to develop and implement the Master Plan, acknowledged a clear gap in its planning efforts. In 2013, the CPRA initiated a community focus group, as well as a landowners' focus group, to provide input into the continued evolution of the Plan. The CPRA invited tribal leaders and involved tribal citizens from coastal Louisiana to participate in the community focus group. However, accepting such invitations is challenging, as there is no compensation for travel to participate in the meetings in Baton Rouge, which is over 100 miles from the tribal communities, or support for child care or time away from an overabundance of co-occurring issues the leaders have to juggle and address on a daily basis, on top of their own work and family commitments.

After figuring out the logistics of time off work, childcare, and time away from other pressing tribal affairs, Chief Shirell (Grand Caillou/Dulac) and Theresa (Pointe-au-Chien) agreed to attend the first focus group meeting in Baton Rouge in July 2012. Yet, instead of spending the time discussing actual protection and restoration issues for their communities, they were sidelined with having to take the time to go through the maps they were handed and ask why, if they were invited, their communities were still not included on the maps; the maps were labeled as if only one larger tribe inhabited the region, instead of different distinct tribal nations, with their own needs, resource concerns, governments, and membership. While given the gesture of a seat at the table, their time was usurped by still having to prove who they are.

It appears that the Tribes' exclusion from state-led mitigation and restoration efforts will continue as funds from the Resources and Ecosystems Sustainability, Tourist Opportunities and Revived Economies of the Gulf Coast States (RESTORE) Act[20] are distributed. The RESTORE Act requires 80 percent of the Clean Water Act penalties paid by the parties responsible for the 2010 BP Deepwater Horizon Oil Disaster to go toward Gulf Coast restoration. However, the state of Louisiana has dedicated all RESTORE Act funds to be spent on projects contained within Louisiana's 50-Year Master Plan for a Sustainable Coast, which has mostly left out the three tribal communities. While the Gulf Coast Ecosystem Restoration Council notes the importance of including tribal input in restoration activities,[21] the plan to restore the Gulf Coast only acknowledges affected tribes as those that are federally recognized.[22] This excludes the three Tribes because they have thus far been denied tribal federal recognition.

★ ★ ★

Despite the layers of disasters and risks, most communities are focused on staying in-place, and working to restore their lands and maintain their cultures. However, under extreme circumstances, such ecological changes and lack of protection can force community leaders into the difficult decision of whether the community needs to relocate to continue their community and culture, and, for tribal communities, their rights to sovereignty, as outlined in the United Nations Declaration on the Rights of Indigenous Peoples (UNDRIP).[23]

As a last resort and at increasing risk of extreme weather events and habitual flooding, the Isle de Jean Charles Tribal Council recognized they are out of options for in-situ adaptation for Tribal members residing on the Island, particularly after the Island was cut out of the state-sponsored hurricane protection system, and has made the difficult decision to pursue community resettlement. The Council has prioritized the need to maintain the Tribe's community and culture, even if doing so means resettling together in a new location. The following chapter describes the process that led to the Isle de Jean Charles Tribal Council's decision, and how the community is now looking to be the first community in modern times in the lower forty-eight states to relocate together as a whole community.

Acknowledgment

A special note of gratitude to Theresa Dardar for contributing her story "I came and I stayed" to this book and for her friendship and local guidance.

Notes

1 "I Came and I Stayed," digital story by Theresa Dardar, Pointe-au-Chien, audio recorded on August 20, 2012.

2 Digital story, audio recorded on August 20, 2012.
3 US Census Bureau (2010a).
4 Cultural re-invigoration is happening not only among tribal communities in coastal Louisiana but also among other minority populations with long historical ties to the region, such as the Cajuns and Isleños (Canary Islanders).
5 Stoffle and Evans (1990).
6 Ibid.
7 Kachko et al. (2015).
8 Conversation with Renée in Isle de Jean Charles, Louisiana, May 25, 2012.
9 Conversation with Michele in Pointe-aux-Chenes, Louisiana, December 20, 2012.
10 Conversation with Co-Chairman Donald Dardar, Pointe-au-Chien, Louisiana, January 17, 2012.
11 Conversation with Nicholas in Pointe-au-Chien, Louisiana, May 4, 2012.
12 Peterson (2011); Maldonado et al. (2013).
13 Conversation with fellowship participant from Cordova, in Grand Bayou, Louisiana, March 6, 2012.
14 Powell and Maldonado (2017).
15 Conversation with Madeleine in Montegut, Louisiana, December 17, 2012.
16 Conversation with Pierre and Renée in Isle de Jean Charles, Louisiana, May 25, 2012.
17 CPRA (2012).
18 Conversation with Chairman Chuckie Verdin in Montegut, Louisiana, April 4, 2012.
19 Burley et al. (2007: 348); Bethel et al. (2011).
20 US Department of the Treasury (2012).
21 Gulf Coast Ecosystem Restoration Council (2013: 1).
22 White House (2010).
23 United Nations (2008).

7

COMMUNITY RESETTLEMENT

Obstacles, Challenges, and Opportunities

A lot of people didn't leave by choice. A lot of people were forced, they had no other choice. They had to leave . . . if they relocate, they will have people come back because it's our culture to be together. It's instilled in us. We have family reunion, powwow, everybody comes back. It's in our nature to be together. The core of those people who ran away from the French and settled there, they were together, close knit to stay together from the whole journey and settle there together. Our people are calling us back. I think it's time we get back to our culture. And unfortunately it has to be a new place, but it could still be done. Alternate plans if relocation doesn't work? You want the truth of Plan B? The cold, ugly truth? If we don't relocate we lose our culture, we lose our, we lose the Island's gone in the next hurricane or the next ten or fifteen years the Island's going to be gone, people will be spread out all over. . . . You will no longer have a place for everybody to come back to.

—Gabrielle, Isle de Jean Charles, relocated to Houma[1]

I placed two aerial maps on a table during a story circle in Isle de Jean Charles in January 2012. One map showed the Island's landmass in the 1950s, and the other in 2011. Looking at the two maps, Pierre said, "I saw the map gonna happen in 2050. See the Island over here, just gonna be no more than a little dot with a pencil, Pointe-au-Chien ain't gonna have nothin' left over there either. Montegut just lil' dot lil' bigger than the Island. . . . And then that part and Dulac and all, Dulac gonna be gone, and west of Houma. . . . All of those parts, that marsh land, that's gonna be all water. That's the map in 2050." Pierre was referring to the map in Louisiana's 50-Year Master Plan for a Sustainable Coast, discussed in the previous chapter, which left Isle de Jean Charles out of the hurricane protection zone.

The Coastal Master Plan had the option of voluntary relocation for individual households (CPRA, 2012), but relocation as a whole community was not included. Henri, an Isle de Jean Charles Tribal Council member, asked at the January 2012 public forum held in Houma for the Coastal Master Plan, "Where in your report is your plan to relocate our Tribe? We're running out of chances for federal recognition. . . . What's the plan to keep our community together? Oil companies came in and left us like dust. Now you want to scatter us like dust."

With little land left or protection from increasingly extreme weather events and flooding, and faced with potential cultural genocide, nearly two decades ago the Isle de Jean Charles Tribal Council made the difficult decision to resettle to another site *as a Tribe*; to maintain their culture, way of life, traditions, and autonomy in a new place; and to bring their community who has been forced to scatter back together.[2] Chief Albert said he liked to think of relocation as restoration, as that seemed like the only option they had.

Yet, over the years the Tribal Council has run into countless policy obstacles and practical and regulatory challenges of preventive community-led and community-wide resettlement. There is a lack of government policies and measures supporting *whole community*—as opposed to *individual*—resettlement. For example, the US Federal Emergency Management Agency (FEMA), under the US Department of Homeland Security, is the primary federal agency designated to respond to, prepare for, and mitigate disasters in the United States. FEMA's Voluntary Buyout Program is designed to support homeowners that are experiencing repeated flooding and would like to relocate. FEMA provides grant-funding support to local and state governments, which initiate and administer the buyout projects. However, the Buyout Program only supports *individual* relocation. Organized on an individual basis—as opposed to a whole community resettlement—the government disaster programs do not consider the disruptions to culture, sense of autonomy, and community values, including the place-based communal and environmental relations that tribes and other place-based communities value, that can occur when people must move in a piecemeal manner.

There are no funds for the lengthy and sensitive process of resettling an entire community together and there is no federal government agency delegated to support communities' proactive efforts.[3] Without such support, when plans fall through, communities are forced to start the process all over again, losing time that they can no longer afford. This chapter discusses the obstacles, challenges, and opportunities the Council and engaged tribal members have experienced in working toward proactive community resettlement.

★ ★ ★

Finding Home

By Babs Bagwell, Isle de Jean Charles tribal member[4]

For as far back as the mind allows me to recall I was told that I was

> Spared . . .
> Saved . . .
> White.
> Adopted,
> Not Relocated.
> They were wrong.

When busing of African Americans began in Louisiana
And the fights broke out
I was told by the principal of the school to stay home for my "safety"
I suppose it was due to the color of my skin being not African-American
Nor White.
For all legal purposes I WAS White
AND terribly ashamed of what the White man was doing.

It has been a life-long journey to find my TRUTH.

I had been running in circles around the home of my people.
Knowing but at the same time not knowing I was on the edge of one of the
    greatest discoveries of my life.
The light, the smells, the sounds there called to me
Like a drum beating in the distance
And I listened.

After years of asking questions and another year of extensive research
I found myself not only smack dab in the middle of the circle I had been
    running in
But also, with my feet upon, the MOST BEAUTIFUL place on earth.
Isle de Jean Charles, Louisiana,
Home of one of the Biloxi-Chitimacha-Choctaw tribes of Louisiana
And,
My ancestral home.

Surrounding me was all the light,
The sounds,
The smells that had drawn me to this place.
Before me,
The fragments of what was left to become my ancestor's fate if action was
    not taken.
Confirming to me that beauty is not what the eye alone beholds,
But what the heart feels when the eyes see.

Despite the distance I had been forced to experience
I felt knowledge of the sadness
Knowing what the possibility of relocating from this place,
Would force my people to endure.

We WERE a people who led a self-sustaining life.
We fished, we trapped and we gardened some of the most fertile ground
    on earth
Not only for income,
But to provide nourishment for our people.
We were a model for how to live when the White Man first arrived,
NOW we have become expendable in their quest for more.

First the White Man took our women and tried to make them White,
Next they took our lands for their own profits.
They have cut thru our marshes,
Our ancestral mounds,
And left our once fertile lands, barren from salt water intrusion.
They have poisoned our bounty from the waters with their quest for
    monetary gain from oil,
Not only with the oil itself but from the chemicals they used to cover up
their mistakes.
They have killed our trees which were once plentiful and marked our lands
And left in their place a shadow of once was.

They offer relocation but not really,
The election itself had been set up to fail.
People have to relocate themselves so that they may live
Or for those who stay . . .
The reality that they may sink with the Isle OR
Be living on a houseboat above the land that once existed.

Our children who once were surrounded daily by loving ancestors
Now are surrounded by the White Man's ways and strangers.
They know not the power that comes with the beauty of being surrounded
    by your people,
Your traditions.

Separation and Relocation
A good thing?
I can definitely say NOT.
One can be separated from their people and their lands
But the heart . . .

Always knows HOME.

FIGURE 7.1 Canals cut through the marsh near Isle de Jean Charles.

Source: Photo courtesy of Babs Bagwell, 2012.

FIGURE 7.2a, b, c   Isle de Jean Charles landscape.

Source: Photos courtesy of Babs Bagwell, 2012.

★ ★ ★

The Tensions of Resettlement

> We are presented with a [Coastal Master] plan that contains no equitable balance . . . the only option our people have had is to relocate . . . doing so individually will annihilate the beauty of what it is to be an American Indian resident on the Gulf Coast.
> *Chief Shirell Parfait-Dardar, Grand Caillou/Dulac*[5]

Chairman Chuckie (Pointe-au-Chien) voiced concern over whether people from Pointe-au-Chien would be able to stay in their current location in the long-term; however, Pointe-au-Chien is now included in some hurricane protection projects and restoration efforts, and is adamantly focused on staying in place. Explaining why the Tribe is not going anywhere, he spoke slowly and quietly, but deliberately, "You can't just move people from the only way of life they've ever known. People here are fishers, that's all they've ever known."[6]

The Grand Caillou/Dulac Tribe is continuing to work on restoration efforts, but, after witnessing the number of years it is taking Isle de Jean Charles to work on re-establishing the community in a new location, the tribal leaders would like to have a plan in place in case they need to consider the last resort of resettlement. Given the lack of hurricane protection and rapid speed of land loss, erosion, and increased impacts from storms, they want to be prepared. The leaders are trying to figure out the best plan of action because they do not want to relocate as individuals. If it comes to the point where no other options exist to stay in-place, they would like to resettle as a community to preserve their culture and heritage and be able to continue their efforts for federal recognition.

Community leaders of at-risk coastal communities face the challenge of doing what they can with limited time and resources to maintain their communities in place as long as possible. Concurrently they recognize that community-wide resettlement plans take many years to develop and carry out, and cannot be thought about once the floodwaters are upon them. However, if they come to the decision that the community is too at-risk to continue to adapt in-place and decide to pursue resettlement, the community could be cut off from funding to support their current existing infrastructure. The lack of physical infrastructure and ensuing environmental degradation exacerbates deteriorating conditions with no known resettlement date in sight, leaving residents in a perpetual state of limbo and furthering impoverishment, health, and sociocultural risks.[7] For example, the villages of Newtok and Shishmaref, Alaska, voted to relocate in 1994 and 2002 respectively. They have been cut off, by and large, from state and federal funding for much needed infrastructure once they declared the resolve to relocate, but have still not been able to move due to financial and political barriers.[8]

Similarly, for Isle de Jean Charles, the gas company said that it was not worth repairing the gas line when it broke, and instead switched people to butane tanks. By not repairing the gas line, the message received was that the Island community was not worth investing in. Some Isle de Jean Charles Tribal members raised concerns about what would happen once the aging water line goes, or the only road connecting them to the mainland is damaged from another storm; it took years for the road to be repaired after flooding from Hurricanes Gustav and Ike in 2008, and there was uncertainty if the road would be repaired again.

Acknowledged Injustice

The Morganza-to-the-Gulf of Mexico Hurricane Protection System, a flood control project crafted by the US Army Corps of Engineers (USACE), the Louisiana Department of Transportation and Development, and the Terrebonne Levee and Conservation District to reduce hurricane and storm damage in coastal Louisiana, was originally designed to include Isle de Jean Charles within the proposed levee alignment. This meant the community would be protected from flooding. The reconnaissance for Morganza started in 1992 and feasibility started in 1995.[9] However, in 1998, the USACE decided it was more economically feasible to relocate the people on Isle de Jean Charles than to build the levees around the Island and include them in the protection zone.[10]

The Morganza Environmental Impact Statement (EIS) acknowledged that leaving Isle de Jean Charles out of the proposed levee alignment and the likely induced flooding during storm events when the protection system is closed is a potential environmental justice issue. However, the EIS went on to state, "Providing hurricane risk reduction for Isle de Jean Charles has been determined in previous Corps of Engineers analyses to be cost prohibitive."[11] A preliminary nonstructural plan was developed to reduce risk to people and structures that were located in high-risk flood areas. The EIS reported that impacts to the communities left out of the system "would be mitigated through 100% buyout and uniform relocation assistance."[12] However, the offering of support for *individual* relocation only further scatters communities, tears apart social fabric, and can lead to the loss of cultural practices—and the culture itself—that connect people together. The process to relocate individually is colonizing in and of itself and perpetuates the colonial-driven forced assimilation policies.

Local Knowledge Ignored

Chief Albert described going to a USACE meeting in Bayou Dularge after Isle de Jean Charles was cut out of the Morganza project in 1998. The USACE representatives told him, "The realignment was probably not going to be done because they said there's a soft spot from over there where they surveyed at, but there wasn't a soft spot. So they told us if we could find the ridge they would reconsider. So we went and showed them where the ridge was."[13]

The USACE informed the Tribe that if they could find an appropriate ridge to build the levees on, they would reconsider including the community in the hurricane protection system. However, Henri, an Isle de Jean Charles Tribal Council member, explained that when people came to take soil samples for the Morganza project, "if they had listened to the elders and went where they said, they would've found a ridge because there's a ridge that runs up through there. They said after doing a soil sample, cost-ratio it wasn't worth it . . . if they'd listened to the old people they would've found what they was looking for."[14]

I asked Jean, an Isle de Jean Charles elder, what had happened with the Morganza project. He explained that after Isle de Jean Charles was cut out of the hurricane protection system, he wrote to the officials and told them,

> If you come from Lower Terrebonne, it's almost a straight line could come and build a levee and take the Island and go connect with the Lafourche levee system and you could save miles and miles and miles of marshes, but they said it would cost too much. The soil would not be able to sustain a levee, so they moved it close to Pointe-au-Chien over there and we're left out. But at least they gave us a ring levee.[15]

While the earthen ring levee around the Island kept some flooding out of people's yards, many felt that, based on where the levee was placed, it was really put there to protect the Louisiana Department of Wildlife and Fisheries' land on the north side of the Island (Figure 6.8).

Jack, a Cajun elder from upper Pointe-aux-Chenes, told me that when he was on the Terrebonne Parish Council he tried to fight the USACE cutting Isle de Jean Charles out of Morganza: "I said how in the world you can put a cost-ratio on sentimental value? People is buried there. I said how much your place where your family is buried, how much is it worth to you? You tell me."[16] The USACE maintained instead that the cost-benefit ratio was not there to include the community in the levee system.

The Atrocities Continue: The Injustice of Cost/ Benefit-Based Exclusion

The USACE former Morganza project manager, Rodney Greenup, explained that the hardest part about the decisions that went into the Morganza project was "talking to locals like Isle de Jean Charles that can see the levee but can't participate." The community was cut out of the hurricane protection system because the "economics just aren't there." The USACE is "federally allowed to flood some [communities] and protect others," they just need to "justify the cost to save one community over another."[17]

Jerome Zeringue, the former Director of the Terrebonne Levee and Conservation District and current Executive Director of the Coastal Protection and Restoration Authority (CPRA), explained that the project would protect over 90 percent of Terrebonne and Lafourche Parishes. The only real tweaks made in the plan were no longer including Isle de Jean Charles and lower Dularge because of cost-benefit analysis.[18]

The cost-benefit analysis used to make coastal restoration decisions does not account for the distribution of costs and benefits or important social and cultural factors, such as people's identity, beliefs, traditions, livelihoods, and sacred places. It does not include the social, psychological, and financial costs associated with moving fishing families inland, loss of local knowledge, and the mental stress of being removed from one's home and traditional life. These results stem, in part, from a lack of policies to fully include tribal or traditional knowledge or values in project cost-benefit or impact analysis.[19]

Considering such decision-making, anthropologist Clinton Westman (2013: 112) found, in an analysis of impact assessment documents for Canada's Alberta tar sands, that at "the root of these discussions lie differentials in power: power to tell the story of the future and then to enact it." Such processes reflect and reproduce social inequalities that turned coastal Louisiana into an energy sacrifice zone—and sacrifice its communities along with the land. The cost-benefit based restoration and mitigation decisions, which are legitimized by government authorities, need to be critically scrutinized. They dictate and determine who is being sacrificed for, what author and political activist Arundhati Roy (1999), described as the greater common good; the concept of "good" is predominantly based on economic measures, discounting the non-material components that enable communities to function and thrive.[20]

Uprooted Shock

Rodney also explained that levee districts could expropriate people's property, meaning they take someone's land and give the person compensation. He followed this explanation by saying, "and this is America,"[21] pointing to the hypocrisy that plays out in a proclaimed democratic country, where all people supposedly have equal rights. He also raised the issue of eminent domain, in which private property is expropriated in the name of public interest for the greater good. The question becomes how to translate who is included in the public interest and who determines what is in the public interest.[22] Furthermore, compensation alone ignores the negative social impacts wrought by forced displacement, such as loss of social networks, does not account for the time-gap from disruption and recovery, and often results in those displaced suffering from increased impoverishment and marginalization.[23]

Researching the impacts of urban renewal around the United States, health expert Mindy Fullilove (2005: 4) found that through forced displacement, people experience "a 'collective loss' . . . the loss of a massive web of connections—a way of being." She described such loss as *root shock*, defined as "the traumatic stress reaction to the destruction of all or part of one's emotional ecosystem."[24] Root shock at the level of the individual is "a profound emotional upheaval that destroys the working model of the world that had existed in the individual's head. . . . Root shock, at the level of the local community . . . ruptures bonds, dispersing people to all the directions of the compass. Even if they manage to regroup, they are not sure what to do with one another."[25] Thus, no amount of compensation can replace what has been lost.

Implicit in relocation is that residents have to go somewhere else, and those other places can be blocked by host communities' feelings toward the potential new residents. Thus, the question remains where it is that those needing to relocate can go. Social disarticulation becomes more severe when people are relocated farther away and the informal structures they rely on, such as exchange practices and social networks, for their traditional risk management system are scattered and shattered.[26] However, the higher ground near the tribal communities was mostly gone, grabbed by multinational oil and gas corporations and private land developers, and with the area projected to face among the highest rates of relative sea level rise worldwide,[27] the communities forced to move will have to go even farther inland, competing for the same job opportunities and land with host communities and others relocating, which could lead to conflict and increased social tensions.[28]

Failed Options for Relocation[29]

After the Island was cut out of Morganza in 1998, and without government support to mitigate the flooding and restore the land, the Isle de Jean Charles Tribal Council informed the USACE that they would like relocation assistance but as a community. The USACE worked with them to identify a site nearby where the community could rebuild. The USACE hired architects for the relocation proposal with the idea of maintaining a cohesive community to be consistent with the Tribe's federal recognition process. But Rodney Greenup noted that when it came time to vote to determine if the Isle de Jean Charles Tribe would be offered relocation, the majority of people from Isle de Jean Charles did not want to relocate.[30] However, the USACE, not being aware of who were members of the Tribe, counted non-residents and members of another Tribe in a negative "community" vote.

Jerome Zeringue explained how some people said they did not want to leave because they thought that "Big Brother" was going to take their land and drill for oil.[31] Some residents perceived the government wanting them to relocate so the oil industry could have free range over the area without interference. Individuals from local tribes grew up either with the personal experiences or with the stories of tribal and family lands being confiscated by oil and gas corporations, land speculators, and foresters. The USACE representatives and other involved parties failed to recognize the sensitivity of government representatives raising ideas about relocating a tribal community and the continued historical trauma from prior forced removals.

Henri described what he said to the agency representatives at the meeting,

> The only thing you're doing, you're re-living the Trail of Tears. . . . I said all you've done, you're gonna build this levee right there and we're gonna have a

strong hurricane with tidal surge. I say it's gonna come and wash our people to the levee. . . . So I said what you're doing instead of leaving us on the wayside, you're just dragging us out to sea.[32]

In 2009, making another attempt for relocation, Chief Albert spoke at a Terrebonne Parish Council meeting about raising funds to purchase an available property in Bourg, approximately 20 miles north of the Island. But after less than five minutes the Council silenced him. A Parish Council member stood up and raised the concern of property values decreasing if the tribal community moved to Bourg, a predominantly Anglo-American community. Theresa (Pointe-au-Chien) had attended the meeting and described how the Council member stood up and said what would happen to the property value if "those people" moved in.[33]

Driving around the Island together in early 2012, Chief Albert commented,

> They probably want this island to diminish because we're moving into other communities and so the kids that we have will marry into the community and eventually the Indians are wiped out. Ask in Pointe-au-Chien and Dulac, wherever the Indians are at, they're going to move into these other communities and well, south Louisiana won't have any more Indians. So yeah, I think that Andrew Jackson is going to get his way. He's going to wipe out the Indians. Those that will still exist will be those that are federally recognized because they have their little reservation. Our reservation here is the one we have. We moved here so we wouldn't be captured by the Whites and sent to Oklahoma.

He laughed, "I wondered what they'd do if we asked to move to Oklahoma," referencing back to the forced removals during the Trail of Tears.

Paradise Elsewhere?

In May 2012, I sat with Chief Albert in his house in upper Pointe-aux-Chenes, listening to his visitor, Joe, a Native man from outside Louisiana propose ideas for the BP settlement and a relocation scheme. Joe explained that a private developer had bought thousands of acres of land near the coast of Mississippi and was willing to sell it to the Tribe. A couple hours later after Joe left, Chief Albert and I stood together on his front deck. His hands rested on the railing. He sighed, "It's just so far from home."

At Chief Albert's request and with the Tribal Council's consent, a few weeks later I drove 170 miles northeast to Mississippi to meet with the developer and see his plans in person. I had an Isle de Jean Charles Tribal member who lived in Mississippi meet me there and together we walked into the developer's high-rise condominium looking over the Gulf water. Joe was also standing there, claiming to be in town on other business. We sat around a large dark wooden table ready with four chairs. I slid in but the chair slanted me back. I scooted upright to be eye-level with the two men. The walls were covered with maps of development projects the developer had underway, including a theme park, where he said tribal members could seek employment.

Joe and the developer drove us around the 1,000-acre property the developer was proposing to sell to the Tribe. The property ran parallel to a river a few miles inland from the coast. On one side of the property was a development with McMansions[34] and the other side was trailer parks. They then drove us to City Hall to meet with the Mayor. The developer talked about how there were state politicians from Mississippi that wanted the Tribe there and would back them, just that, he joked, they could not burn the ships this time, referencing the Biloxis actions when they were forcibly removed a couple centuries prior. The Mayor felt that the City would not want the Tribe there because it would fear interference with the city's gaming industry if the Tribe put in a casino. I looked over at the developer and Joe, both of whom seemed to have business connections with the gaming industry in the area.

After our good-byes, I drove back late that night to our camp in Pointe-au-Chien. I thought of how Chief Albert joked about putting a sign up at another possible relocation site inland near Houma to say, "Isle de Jean Charles 2." Later that same night, I dreamt about the new potential site, which the Tribe did not yet have funds to purchase. In my dream the land was right along the bayou, filled with big cypress and oak trees. When I awoke, I realized that the images in my dream were from pictures I had seen of what the Island used to look like. The Chief's words echoed in my ears, "It's just so far from home."

"A Symptom of What's Happening Everywhere"

Isle de Jean Charles is experiencing some of the most rapid land loss in Louisiana and the world, but it is not the only community in Louisiana at-risk of displacement. Descendants of other population groups who have been in coastal Louisiana for centuries, such as the Cajuns, Isleños, and African and Caribbean groups, also face similar risks. The 2017 Louisiana Master Plan for a Sustainable Coast highlighted that over the next 25 years, ten communities along coastal Louisiana—including the three Tribes—will be particularly vulnerable to flooding, and 11 communities—also including the three Tribes—are expected to be dramatically changed by flooding in the next 50 years if action is not taken.[35] However, this calculation does not account for effects from future hurricane damage and is based on a medium environmental scenario for sea level rise, which evidence shows we are far surpassing.

Kerry St. Pé, Executive Director, Barataria Terrebonne National Estuary Program, articulated what these numbers actually mean: "People are moving away. Farther up and farther out, out of state. And that's the one thing I fear most because that's the culture, everything's about that. The fact that we've kept these people here generation after generation and we're losing that. Isle de Jean Charles is just a symptom of what's happening everywhere."[36]

Indeed, Isle de Jean Charles and other coastal Louisiana communities are not alone. A number of coastal communities around the world are threatened by displacement and the loss of livelihoods, social structures, cultural practices, traditions, and ways of life. However, instead of waiting until policies or legislation are passed, some communities are exercising their agency and proactively pursuing community-led relocation to ensure not just their survival, but their sovereignty, rights, and integrity. Below I highlight examples of Indigenous coastal communities pursuing proactive relocation initiatives.

In the Arctic, climate change is having clear impacts on the landscape including diminished sea ice and increased windiness, storminess, and coastal erosion. Public infrastructure has been significantly damaged or destroyed due to the combination of extreme weather events and on-going erosion. In particular, the Native villages of Kivalina, Shishmaref, and Newtok have been working to relocate for decades, echoing the experiences of Isle de Jean Charles. A number of residents fear that without an organized relocation a large flood would result in diaspora during or following a disaster event, removing their community from traditional subsistence territory and severing human-ecological relationships that have persisted for thousands of years.[37] Repetitive flooding events have led community leaders to look to relocation as a long-term solution to risk.[38] While Newtok has come the farthest in its efforts and relocation has begun in initial stages at a new site location, the Tribal Council has faced policy and funding barriers at every step.[39]

Some tribes in the Pacific Northwest whose tribal residents, infrastructure, and facilities are at-risk of climate-induced increased flooding, storm surges, sea level rise, and tsunamis are also proactively pursuing relocation for the villages and infrastructure in danger. However, relocation has become a deeply political process that connects back to land grabs from previous centuries when tribes lost much of their ancestral territories through treaty processes with the federal government. For example, Congress passed legislation in 2012 to transfer 785 acres of Olympic National Park back to the Quileute Tribe in northwest Washington to provide for tsunami and flood protection in their efforts to relocate some of their village homes and buildings to higher ground; without full funding, however, implementation remains a challenge.[40] The Quinault Indian Nation has developed plans for relocating its lower village of Taholah to approximately 200 acres of higher ground one-half mile from the existing village within a culturally connected location. The Nation still needs funding and institutional support to fully implement its relocation plan.[41]

A number of Small Island Developing States pursuing proactive relocation are facing particular citizenship, sovereignty, legal, and economic concerns; many low-lying islands and atolls are only a few feet above sea level and have limited options for internal migration.[42] For example, the Carteret Islanders in eastern Papua New Guinea, facing extreme erosion and immediate threat of king tides, saltwater intrusion, and sea level rise, formed a Council of Elders in 2006, which led to the development of the Carteret Integrated Relocation Project to organize the voluntary relocation of Carteret Islanders approximately 60 miles northeast to Bougainville Island. The first group began to relocate in 2009.[43] The Guna people of the San Blas archipelago in the Caribbean are pursuing relocation options due to sea level rise and limited space for a growing population. With funding from the Inter-American Development Bank, the community on the island of Gardi Sugdub made arrangements to obtain 17 hectares of land on the mainland; they recently paid with their own funds to have the land cleared to prepare for relocation.[44] Kiribati has established a "migration with dignity" strategy that focuses on education and skills training for youth to be employable if the need arises to migrate to other countries or, in the worst-case scenario, their islands are no longer inhabitable.[45] And the residents of Vunidogoloa village in Fiji, with financial assistance from the government of Fiji, relocated in early 2014 because of encroaching sea level and inundation events that frequently damaged the community.[46]

In all of the above examples, recognizing that they have no or limited options left for in-situ adaptation, communities are actively leading initiatives for their relocation efforts to culturally significant (or as nearby as safely possible) sites of their choosing that maintain their community's rights and cultural integrity and sovereignty, as outlined in the United Nations Declaration on the Rights of Indigenous Peoples (UNDRIP, 2008). Across these varied contexts common focuses include maintaining or gaining economic self-sufficiency, food security, and access to traditional resources, moving together instead of individually, and whenever possible, staying within ancestral territory, raising the concern that pushing people into urban locales and away from their traditional lands threatens cultural genocide. Resources are still needed in their current locales to protect people's health, safety, and well-being between now and when relocation is actually carried out. Also in common is the lack of established institutional arrangements for full implementation of a sustainable relocation process. Fiji is an exception, with the Fiji Government in the process of developing national guidelines, but thus far the process appears to be a top-down initiative without participatory planning.[47]

Some major concerns include how communities will be able to make decisions about relocation, maintain their communities and culture, and reduce the impoverishment risks that often go along with relocation without funding mechanisms or governance frameworks for support.[48] Governments are either inconsistently implementing social protection policies or failing to create such policies at all, perpetuating a state of limbo for communities who have been trying to proactively relocate, some for over a generation. There is currently no institutional framework at the nation-state or international levels governing environmental disasters that occur incrementally and require relocation: there is no policy to determine whether, when, and how climate-influenced relocations should occur.[49]

Isle de Jean Charles is among the communities leading the way in overcoming institutional barriers. The Tribe is out of time to wait for legislation and policies to be worked out. Faced with immediate threats of cultural genocide and health and safety concerns, and knowing that once the water overtakes their land it is too late, the Isle de Jean Charles Tribe is continuing to move forward to implement its plans. As an ally supporting the Tribe's efforts noted at a Congressional forum (EESI, 2015), "They are not going to be climate refugees, these folks are Indigenous scouts for a new way in the 21st century to be able to thrive with culture, values, and dignity in tact."

Moving Forward with Community Resettlement[50]

The Isle de Jean Charles Tribal Council, in partnership with the Lowlander Center and a team of volunteers, including community organizers, researchers, developers, architects, engineers, and practitioners, has developed a community-based, culturally appropriate, renewable energy-driven resettlement plan, including ecosystem protection and rehabilitation of the Island to mitigate further degradation of their current home and ancestral lands, to once again be a self-sustaining community.[51] The plan has several key innovations. One important aspect is how *community* is defined. The plan is not written for Isle de Jean Charles, the geographic location. It is designed for the Isle de Jean Charles *Tribe*, recognizing that the majority of tribal members have been forced to leave the Island over the years due to habitual flooding, accumulating disasters, loss of livelihoods, and co-occurring adverse events. The plan is envisioned to restore the *whole community*, which includes providing the option to people who have already been forced to move the ability to come back together. A number of people who already relocated are still within the greater area, wanting to stay as close to their family and cultural ties as possible, and have voiced the desire to come back together if given the option.

Recognizing the continued importance of the Island even if people move—with access to traditional fishing places, a cemetery, sacred mounds, and a common place that ties people together—the plan includes working to restore the Island. Even if it is eventually no longer habitable, the lands should be protected and maintained under the stewardship of the Tribe as part of its territory. The plan was guided by the vision of the best principles in resettlement, accrued from decades of documented past resettlement experiences from around the world, and to honor the Tribe's cultural values throughout the resettlement process. Understanding that resettlement does not end after houses are constructed and people move in, the plan includes a vision for the various types of infrastructure—built, economic, social, natural, cultural—that are required to successfully re-establish a community.

To move forward, with very limited resources, the tribal leaders have been vocal in high-level government discussions—and created their own governable spaces—engaging with policymakers, the media, non-profit and non-governmental organizations, practitioners, scholars, and students from around the world. These efforts have now proven results: in January 2016, the US Housing and

Urban Development Agency in partnership with the Rockefeller Foundation announced a $48 million grant award from the National Disaster Resilience Competition to support the Isle de Jean Charles Tribe to resettle as a community. If seen through, this project would make them the first community to resettle together, as a community, in the continental United States in modern times. The award was granted through the state of Louisiana, which received $92.6 million in grants through the Competition for both the Louisiana Strategic Adaptations for Future Environments (LA SAFE) Fund and the Community Resettlement of Isle de Jean Charles.

The Tribe, along with the Lowlander Center team, was able to achieve this with expertise of tribal members and subject matter experts, all based on a volunteer basis and all of whom have contributed personal funds, resources, and time. The lack of funds becomes a hindrance for the team to access resources or to be present at essential meetings and conferences. Furthermore, the Tribe is constantly dealing with on-going issues, such as flooding and disaster recovery. Resettlement planning and fundraising are in addition to the workload demanded by the chronic and daily disasters.[52]

At the time of this writing, it is not yet known how the awarded funds will actually be spent for the Tribe's resettlement. This is due in part to conflict between individual-based federal policies and the necessity of the Tribe to resettle as a whole community. For example, the Fair Housing Act,[53] which protects people from discrimination to secure housing, is important and honorable in terms of inclusion, but it also poses a distinct challenge because it could open the new resettlement site to those outside the Tribe, when the Tribe's plan was about regrouping the tribal community together. Most federal policies and programs, such as FEMA's Voluntary Buyout Program, are written and designed based on *individual*—not *community*—rights, in effect furthering colonial-based policies in conflict with the Tribe's community-based values. In the case of resettlement, these policies do not provide the ability to re-establish the physical, economic, cultural, social, and spiritual fabric that enables community well-being.[54]

Further, with the media, agencies, and other entities postulating the Isle de Jean Charles Tribe's resettlement as *the* model for resettlement in the United States, there is real risk for the precedent that could be set for other communities' current and future resettlement needs based on how the Isle de Jean Charles Tribe's resettlement unfolds. If the state pursues a market-based approach turning the resettlement into an individual economic transaction by individual household, there is tremendous risk of cultural genocide, as households could be pushed on an individual basis into subdivisions or urban locales away from culturally-significant places, and lose access to subsistence and cultural resources and practices, social networks, traditions, and ways of life. Such a scheme continues to propagate the colonial agenda, inflicting damage to physical, mental, emotional, and spiritual health, both on individuals and the community as a whole. Some of the risks are readily demonstrated through the way that the Isle de Jean Charles' resettlement is being framed for the public.

Culture of Representation: The Framing of "Climate Refugees"

Media reports following the announcement of Isle de Jean Charles' resettlement funding have over and over splayed the headline, America's first "climate refugees."[55] There are several deep problems with this type of framing: attention is diverted away from the underlying causes of both climate change and displacement, and local residents are framed as agent-less victims. By grabbing a catchphrase and splaying that as the headline—*climate refugees*—representations of communities today can be virtually replaced by those 200 years ago.

Discussing terminology is about more than semantics; labels are important because they "impose boundaries and define categories . . . labeling can shift—or sustain—power relations in ways that trigger social dislocation and prejudice efforts to achieve greater equity."[56] Colonial projects march on, obscured by thin veneers depicting actions communities are doing, but still with the undercurrent of continued victimization; stripping of agency and empowerment; narratives that portray representations of poverty, scarcity, and a culture that is static and lacking.

Further, sociologist and political economist Stephen Castles (2002: 9) explained, "definitions reflect and reproduce power, and none more so than the refugee definition. . . . It makes a big difference whether people are perceived as refugees, other types of forced migrants or voluntary migrants." While the majority of people displaced by climate change impacts will be internally displaced,[57] the refugee definition is based on a person being persecuted and forced outside their country of origin. Thus, the labeling process can create a dichotomy between citizen/foreigner and insider/outsider. The refugee label elicits a context of foreignness, socially constructing members of the Tribe as *the other*, implying that they lack real citizenship.

We need to be mindful of the role the media, policy and decision makers, and researchers play in what anthropologist Gregory Button (2001: 145) cautioned as, "reproducing ideologies and reinforcing the privileged positions of authority." Such discourses propagate the victimization image of those most at-risk of climate-influenced displacement, creating a "state of panic."[58] This often also works to justify the construction of border fences—both literal and symbolic, like the Parish Council member trying to keep the Tribe out of Bourg—while obscuring accountability for responsible parties and systems that have constructed people's vulnerability to such displacement in the first place.[59]

Collective Action

Isle de Jean Charles and other communities such as Shishmaref and Newtok have spent a generation or more working toward relocation. Their efforts have been impeded at every step due in large part to a lack of existing institutional and governance structures to assist their efforts, and funds only made available after a disaster occurs. Faced with immediate threats, while legislation and policies are worked out, these communities are continuing to move forward, reach out for support in partnerships, engage together, and pursue community-led actions.

Due in large part to their efforts, the President's State, Local, and Tribal Leaders Task Force on Climate Preparedness recommended in its 2014 report to the President,

> The Federal Government has an opportunity to provide international leadership by establishing an institutional framework for responding to the complex challenges associated with climate-related displacement. This framework will help Federal agencies and partners provide coordinated, critical support to affected communities across the United States. State, local, tribal, and territorial entities should be consulted and involved in the development of the framework.[60]

This statement echoed the 2013 US Congress Bicameral Task Force on Climate Change report.[61] However, with the current White House Administration systematically deconstructing the progress that has been made on climate policies and actions at the US federal level, including for relocation, it is clear that communities most at-risk need to continue to pursue creative solutions and seek partners and allies across cultural, geographic, and institutional boundaries. Communities' experiences and observations across extended periods of time, and the multitude of historical, economic, and political factors causing rapid environmental change, upon which climate change is layered, warrants immediate action. For communities who have been trying to relocate for ten, twenty, or more years, time is not on their side.

Acknowledgment

A special note of gratitude to Babs Bagwell for contributing her story "Finding Home" to this book and for teaching me to see how this world dances when we are still enough to sit, listen, and observe.

Notes

1 Conversation with Gabrielle in Houma, Louisiana, July 29, 2012.
2 Maldonado and Peterson (2018).
3 Bronen (2011); Maldonado et al. (2013).
4 Digital story, audio recorded on June 1, 2012. Available online at https://www.youtube.com/watch?v=EPJ6AZFt_Us&t=31s
5 Public comment at the January 2012 public forum in Houma, Louisiana for Louisiana's 50-year Master Plan for a Sustainable Coast.
6 Conversation with Chairman Chuckie Verdin in Montegut, Louisiana, December 6, 2011.
7 Marino (2012); Laska and Peterson (2013); Marino (2015); Peterson and Maldonado (2016).

8 Marino (2012); Tom (2012); Marino (2015).
9 Conversation with Jerome Zeringue in Baton Rouge, Louisiana, July 12, 2012.
10 USACE et al. (2013).
11 Ibid.: 5.53.
12 Ibid.: 6.45.
13 Conversation with Chief Albert Naquin in Isle de Jean Charles, Louisiana, January 4, 2012.
14 Conversation with Henri in Grand Bois, Louisiana, March 9, 2012.
15 Conversation with Jean in Isle de Jean Charles, Louisiana, August 2, 2012.
16 Conversation with Jack in Pointe-aux-Chenes, Louisiana, May 25, 2012.
17 Conversation with Rodney Greenup in New Orleans, Louisiana, July 13, 2012.
18 Conversation with Jerome Zeringue in Baton Rouge, Louisiana, July 12, 2012.
19 Maldonado and Peterson (2018).
20 Oliver-Smith (2010: 142–143).
21 Conversation with Rodney Greenup in New Orleans, Louisiana, July 13, 2012.
22 See Oliver-Smith (2009).
23 Cernea and Mathur (2008); Maldonado (2008); Oliver-Smith (2010).
24 Fullilove (2005: 11).
25 Ibid.: 14.
26 Bisht (2011).
27 Marshall (2013).
28 Scudder (2005).
29 'Relocation' is the term often used in the context of climate change and displacement, such as its use by Alaska Native communities working on relocation. The term 'resettlement' is typically used to refer to a long-term process in the development context, and in this book is also used to describe the process of re-establishing an intact community settlement in a new location.
30 Conversation with Rodney Greenup in New Orleans, Louisiana, July 13, 2012.
31 Conversation with Jerome Zeringue in Baton Rouge, Louisiana, July 12, 2012.
32 Conversation with Henri in Grand Bois, Louisiana, March 9, 2012.
33 Cross-community conversation (telephone) between tribal leaders from coastal Louisiana and Newtok, Alaska in Chauvin, Louisiana, June 5, 2012.
34 McMansions are large mass-produced dwellings.
35 CPRA (2017).
36 Conversation with Kerry St. Pé in Thibodaux, Louisiana, July 27, 2012.
37 Schweitzer et al. (2005).
38 Marino (2012, 2015).
39 Tom (2012); Bronen (2014).
40 Papiez (2009); Quileute (2011); de Melker (2012).
41 Quinault Indian Nation (2015); Moldenke (2017).
42 Keener et al. (2012); Lazrus (2012).
43 Bronen (2014).
44 Displacement Solutions (2015).
45 Tong (2014).
46 McNamara and Jacot Des Combes (2015).
47 Ibid.
48 Bronen (2011); Lazrus (2012); Maldonado et al. (2013).
49 Bronen (2011); Ferris (2012).
50 Authorship of this section belongs as much to the core Isle de Jean Charles Resettlement Team that put together and submitted the Tribe's Resettlement Plan to the state, which was awarded funding by the National Disaster Resilience Competition, led by the Isle de Jean Charles Tribal Council (lead, Chief Albert Naquin) and the Lowlander Center (lead, Kristina Peterson), as it does to the author.
51 Laska et al. (2014); Maldonado et al. (2015).
52 Maldonado and Peterson (2018).

53 For further information, see www.justice.gov/crt/fair-housing-act-2.
54 Maldonado and Peterson (2018).
55 See for example, Davenport and Robertson (2016).
56 Moncrieffe and Eyben (2004: 1).
57 Kälin (2008); Kolmannskog (2008: 4).
58 Lancaster (2006).
59 Maldonado (2016a).
60 President's State, Local and Tribal Leaders Task Force on Climate Preparedness and Resilience (2014: 30).
61 US Congress (2013).

8

A CALL TO ACTION

"We Have Got to Slow the Rising Tide"

> Living here is a commitment. You have to do it in spite of the challenges of storms, flooding, distance to everything. But the good outweighs the bad. . . . It's not everywhere that you can be outside your house, have a nice breeze, stay outside in comfort and safety knowing everyone around you. We have got to slow the rising tide. What it was, what it is, that's what keeps me here.
>
> —Chris Brunet, Isle de Jean Charles[1]

The predicted increasing frequent and intense climatic events, such as flooding, are highly likely to directly and indirectly result in increasing movements of human populations, creating new policy and social challenges. We are already seeing evidence of this around the globe. For example, China is currently implementing the world's largest environmental migration project, resettling 1.14 million residents of the Ningxia Hui Autonomous Region for multiple reasons, including, but not limited to, drought, degraded land, industrialization, and poor policies. Already, profound issues abound in the resettlement process, such as loss of land and assets and degraded living conditions.[2] Unless there are radical changes made in countries' and institutions' policies and financial contributions toward resettlement, these are the types of consequences we can expect.

Grounded in knowledge from previous resettlement processes, this chapter offers recommendations to move policies forward that support communities to reduce risk, mitigate harm, and adapt. The recommendations are divided into sections particular for some of the key issues raised in the previous chapters—resettlement, disasters, and climate adaptation and mitigation—but also address broader issues raised throughout the book that relate to contexts across the climate–disaster–energy nexus.

Resettlement Recommendations

Learn From Previous Resettlement Processes

As climate-influenced displacement becomes an ever-widespread problem and the movement of at-risk populations becomes increasingly necessary, lessons need to be learned and translated from previous resettlement processes to mitigate increased suffering and impoverishment. Measures taken to reduce disaster risks can be key lessons for climate-influenced displacement and relocation. For example, due to the limited possibility of mitigation against natural hazards through other measures, with support from the World Bank and the Inter-American Development Bank, officials and institutions in some countries in Latin America undertook resettlement of at-risk communities as the only possible risk reduction measure.[3] While some communities successfully moved to areas at lower risk of hazards, the process highlighted the lack of attention given to the cultural dimensions, social ties, and accountability mechanisms. Lessons learned include: cash compensation alone is inadequate; legacy issues matter when it comes to the relationship and trust between the government and affected communities; it is essential to include cultural dimensions and accountability mechanisms; the need to incorporate resettlement into a larger, comprehensive risk reduction strategy that includes effective land use planning and several resettlement options; and resettlement needs to be treated as a multidimensional process.[4]

The development-caused forced displacement and resettlement (DFDR) process also provides important lessons learned and transferable knowledge for climate-influenced displacement, principally the use of organized planning as a key instrument in the toolkit for resettlement.[5] Over the past four decades social scientific literature has documented the fundamental impoverishment risks and many negative economic, social, and cultural impacts experienced by DFDR affected individuals, families, and communities, including, but not limited to, impoverishment through loss of land, livelihoods, common natural resources, and culturally significant resources; social disarticulation; food insecurity; homelessness; increased morbidity; physical and mental health consequences; psychological trauma; and economic and political marginalization.[6] The DFDR literature also provides strategies for mitigating harm, reducing risks, and reconstructing people's livelihood following displacement.[7] We must learn from the accrued DFDR lessons and knowledge-base, which has clearly identified potential risks caused by displacement, and the need to minimize them and formulate counteracting policies and strategies to turn barriers into opportunities and benefits for those placed in harm's way.[8]

Establish an Inter-Agency Climate Change-Displacement and Resettlement Coordination Program

Similar to DFDR, a core problem of climate-influenced displacement and resettlement is an economic one: who will finance it? In other words, where will the money come from to help those displaced? However, a key distinction between

DFDR and climate-influenced displacement and resettlement is that when some-one is displaced by a development project, it is because something constructive, such as a hydroelectric dam that provides electricity, is created and there is money that has to be calculated in the process to avoid inflicting harm, such as compensation—however imperfect—paid by an agent who takes over assets. For DFDR, there are resources available from the project's governing agencies that can provide compensation.

In the case of climate-influenced displacement there is no development being created that contributes benefits and there is no institution or agency tasked with paying compensation for lost assets. Furthermore, it is more difficult to pinpoint the culprit of climate displacement. In the case of DFDR, there are govern-ment and funding agencies and contractors directly involved. But for climate-influenced displacement and resettlement, where is the finger pointed and who is deemed responsible to pay up? The State has a different type of obligation than in the case of DFDR; for climate-influenced displacement, it becomes more a responsibility of support and assistance, as opposed to direct compensation for lost assets.

The Paris Agreement, in the section on "Loss and Damage," included a call to establish a task force to, among other things, "develop recommendations for integrated approaches to avert, minimize and address displacement related to the adverse impacts of climate change."[9] This was watered down from an earlier draft that included a "climate change displacement coordination facility," which included such responsibilities as providing assistance to organize migration and planned relocation and undertaking compensation measures for people displaced by climate change.[10]

Such recommendations cannot be watered down. An inter-agency climate change displacement and resettlement coordination program—both at the national and international levels—is needed to help facilitate planned resettle-ments. However, such coordination cannot only come from the top-down. The governance frameworks established need to be flexible so that the communities needing the support can access resources and trigger policies to facilitate the pro-cess, but in a way that is culturally appropriate and adaptable to their particular circumstance.

Establish a Legally Binding, Rights-Based Adaptive Governance Framework[11]

Robin Bronen, a human rights attorney and prominent research scientist has led the call to establish a rights-based institutional framework and legal mechanism to support populations displaced as a result of climate change impacts, as such form of movement is involuntary and has been imposed upon them.[12] The human rights of relocated people need to be fully respected. In particular, people displaced face the potential loss of human rights to adequate food, water, and health.[13] The framework needs to include a system of rights and responsibilities that establishes which government agencies are responsible for supporting the different components of displacement and how agencies should work with community leaders in supporting the implementation of relocation plans. The framework should be flexible to account for varying reasons for displacement, such as direct climate change impacts like permafrost thaw, or the combination of factors, such as sea level rise and development-caused land loss. It also needs flexibility to allow for preventive action, such as a community deciding it needs to relocate to stem further damage from increasing slow- or sudden-onset disaster risks.

Create Policies That Include Protection of Socio-Cultural and Human-Ecological Relationships

The evidence accrued over the last half-century clearly demonstrates that without improved policies and actions to actively work to better manage, plan for, and organize resettlements in a culturally-appropriate and effective manner with the full participation and guidance of the affected population, deep-seated impoverishment experienced by forcibly displaced people will persist. Resettlement policies "must improve and protect socio-cultural and human-ecological relationships or communities will suffer."[14]

Resettlement focused on empowering a community to continue their culture, and understanding the complexity and dynamics by which people experience the world, is much more challenging than at first glance. Culture is tangible and intangible; it is a fluid dynamic process always in an evolving state of change,[15] intertwined with a host of economic, political, and social relations and tensions that are constantly altering seemingly stable processes. By simply regrouping people without the thought of these other components in mind can leave people still feeling dispersed and unsure of what to do.[16]

Provide Community-Wide Resettlement Support

A policy framework addressing climate-influenced displacement needs to provide *community-wide*—not just *individual*—support. For many tribes and culturally connected communities, the need to be recognized as a community, and not just individuals, calls for building in flexibility to existing policies to enable providing community-wide support.

Resettlement is about more than just moving people from one place to another, as if the same model could be used in any geographical or cultural context. Resettlement should be just that—re-establishing an intact community settlement that includes the key physical, economic, cultural, social, and spiritual infrastructure that enable the community to thrive in such a way as they determine, in a location that the community chooses and makes sense to them. It should envision a community that can thrive and resemble the positive attributes of the pre-dispersed community so as to reunite those who have been displaced.[17]

Support Community-Led Resettlement That Empowers and Honors Community Decision-Making

Communities should be empowered to make their own decisions on whether migration is a necessary adaptation strategy or not. Resettlement plans need to be community-led because it is not just about building houses and moving people from one place to another, as if the same model could be used in any geographical or cultural context. There are important elements to consider beyond just physical resettlement, such as choosing a site location, housing configuration, maintaining social networks, livelihood opportunities, access to culturally important resources, and creating a plan for sustainable community development.[18]

Community-led resettlement requires including the communities' voices and input in all decisions and developing respectful relationships between communities, government authorities, and all involved entities. This means understanding people's social and cultural values and worldviews, which requires an iterative, participatory process between project implementers and local populations that considers how different frameworks envision the future.[19] It is imperative that the tribal and community leaders who have spent a generation and more working on such efforts are the ones guiding the process to help ensure that the community's rights and cultural sovereignty are held intact.

Provide Support for Existing and Needed Infrastructure in Place

As is evidenced by the case of Isle de Jean Charles, it is taking communities decades to achieve resettlement. Yet, one distinct consequence is that if community leaders come to the decision that the community is too at-risk to continue to adapt in-place and decide to pursue resettlement, they could be cut off from funding to support their current existing infrastructure, left in limbo between the time the decision is made to resettle and when such plans can be implemented, which is proving to take decades without governance mechanisms and adequate support.[20] Thus, resource support needs to continue to be provided to communities to maintain health, livelihoods, and well-being while they are still in place.

Plan for Current and Future Displacements in the Context of a Just Transition to Clean Energy

Given that we are already locked in to a certain amount of climate change, while working on mitigation, we also need to plan for current and future displacements. After all other means to stay in-place are pursued, resettlement can be used as an opportunity to bring communities who have already been scattered due to disasters and other events back together, rejuvenate cultural traditions and practices, build capacity and long-term resilience, and, through the resettlement process, create renewable-energy driven economic opportunities and community planning in a just transition from toxic to clean energy resources.

Disaster-Specific Recommendations

Support Proactive Strategies for Both Sudden- and Slow-Onset Events

The Stafford Disaster Relief and Emergency Assistance Act—the primary legislation for most federal disaster response activities in the United States, especially as related to FEMA—does not include slow-onset environmental changes such as coastal erosion and sea level rise. Further, funding for disasters is only triggered, for the most part, after a disaster occurs. The Stafford Act needs to be amended to enable FEMA and other involved agencies to support communities that are experiencing not only sudden (e.g., hurricane), but also slow (e.g., erosion) onset events and working on proactive measures to reduce risk, mitigate harm, and adapt.[21]

Include Support for Continuing and Re-establishing Subsistence Practices

Disaster policies are often designed to push people who are displaced from rural to urban locales away from their ancestral territories and into a capitalist-based economic marketplace. This distinctly ignores people's "right to be rural,"[22] and the significance of subsistence practices and informal exchange practices not only in economic terms, but also social and cultural value, with such practices often tied to identity, place-connection, community structure, and cultural traditions. Disaster policies need to include support for people to continue their subsistence practices in-place.[23]

Local Knowledge Should Guide Preparedness, Risk Reduction, and Recovery Processes

In disaster-related policies and practices, culture is often treated as tangible, homogenous, static. The diversity of communities and places, on-the-ground actions and networks for how things are actually accomplished, intricacies of local politics and maneuvering, and layers of sociohistorical inequalities are often missing from expert calculations and official frameworks for action. Even when culture is acknowledged by governance structures, such structures all too often fail to acknowledge the internal heterogeneity of cultures or include the ways in which local cultural framings could guide preparedness, risk reduction, and recovery processes. The basic but most important question often goes unasked to the people who actually live there: "What would you do and how?"

Within the context of a changing climate, as communities face increasing disasters, often still recovering from one disaster when the next one hits, this question becomes ever more important. If people spiral down a recovery road that increasingly becomes unfamiliar, they are further alienated by the process, resulting in increased harm. Participating in, and providing leading guidance to, their own recovery empowers communities to take ownership of the process and maintain their rights to reduce risk and adapt to risk as they see most culturally and locally appropriate. The answer to the above question—*what would you do and how*—should guide the long-term recovery process.[24]

Utilize Social Network Analysis to Inform Policy and Practice

Disaster survivors' capacities to cope with disaster impacts can hinge on the strength and range of their social relations—networks, kinship, and other patterns of group relationships that provide both immediate and long-term support independent of disaster aid and agency and organizational support. Government- and agency-led disaster planning, responses, and recovery processes need to account for, include, and leverage community structures and relationships that facilitate information flows and tangible and nontangible recovery needs.[25]

Social network analysis can unveil who (individuals or organizations) are the core nodes in a community or region, where they are creating solidarity and empowering collective action, and where people turn to find resources (e.g., churches, kinship network). It can also aid in understanding how people who have been displaced following a disaster access information and obtain resources when at a distance from their community and kin networks. For example, network analysis could illuminate who people rely on for reciprocity and resource exchange, so if people are forced to move following a disaster event, it becomes clear what non-economic resources they might now be severed from, and, in the reverse, for people who stay what resources they might have lost. Such understanding is crucial when providing disaster support both in receiving communities for those displaced, as well as for people who are able to stay in the affected community, as it explicates what people have *really* lost, beyond economic means, that enables them to survive and thrive.[26]

Cross-Cutting Recommendations for Climate Adaptation

Leverage Boundary Organizations to Facilitate Collaboration and Build Capacity

Boundary organizations can act as hubs to facilitate collaboration between diverse groups or individuals across cultural, geographic, sectoral, political, ideological, and other boundaries, particularly when there are high levels of distrust, disproportionate degrees of power, or unfamiliarity with each other.[27] Effective boundary organizations act as negotiators and translators, and build accountability, trust, understanding, collaboration, and participation between individuals and organizations on both sides of the boundary.[28] Networks that cross boundaries can help facilitate both disaster recovery and long-term adaptation, as an opportunity to build capacity and connect to resources, support, and potential allies in diverse geographic, economic, social, and cultural locations.[29]

Include Diverse Knowledges in Adaptation Processes and Decision-Making

Governments, international bodies, practitioners, and researchers increasingly recognize that traditional knowledges, which often include knowledge about the interrelationship between species and interconnectivity within an ecosystem, are necessary and valuable to inform and guide climate adaptation. Communities understand ecological change based on their relationships to the environment and generational dwelling in place; scientists and policymakers must be ready to engage with communities and make sense of the impacts being experienced from local people's knowledge, wisdom, and perspectives. This includes an understanding that Indigenous knowledge is a science grounded in long-term observations and monitoring. Including traditional knowledges and "indigenuity" into adaptation planning and decision-making would help to democratize and decolonize the adaptation process.[30]

At the same time, no one knowledge system holds all of the answers to the new climate context. Thus, there is a need for diverse knowledge systems contributing to actions, including the development of capacity-building focused collaborations and partnerships between climate and Indigenous scientists.[31] Incorporating diverse knowledge systems and ways of knowing into decision-making processes needs to be done justly and respectfully, so as to not turn the co-design of planning and implementation into co-optation.[32] Careful attention needs to be given to handling and protecting culturally sensitive information and respecting systems of responsibilities.[33]

Address the Root Causes of Climate Change

Disputes between political parties or a struggling economy cannot be excuses for inaction. We cannot view climate change within a vacuum while the causes behind the issue continue to chug along, spewing forth into the atmosphere. The public needs to demand at the local, regional, national, and international levels a shift away from unsustainable resource extraction and burning of fossil fuels and toward a just transition to clean, renewable energy sources. Further, the companies who knowingly work to distort climate science findings need to be held accountable for their share of payment and restoration.

In the process of a transition to renewable energy, the same communities sacrificed by the fossil fuel industry, and by colonial governments, cannot continue to have their lands grabbed to develop renewable-energy driven infrastructure projects to bring energy resources to places far from their homes. A *just* transition needs to be exactly that—a socially, environmentally, and economically just process. If we continue to apply technological fixes grounded in a system that caused the problems in the first place, we are only spinning our wheels if we expect a different outcome. The new path for a just transition calls for a shift from a privatization paradigm to one of shared common resources that values and respects the human—environment relationship. Kathy Sanchez, co-founder of Tewa Women United—a non-profit organization founded and led by Native women—articulated this foundation as one built on a "relational culture."[34] Above all, to achieve climate justice, decolonization must occur.[35]

Funding and Governance Mechanisms Are a Matter of Justice

Providing support to communities that are proactively pursuing resettlement to keep their community together and continue their culture should not be set up as a competition. It is not a matter of figuring out a cost per person or pitting communities against each other in a race for funding resources. Often, the challenges dealt to communities at-risk of displacement are founded on centuries of injustices and accrued unpaid reparations. It is taking decades for communities to move forward with their efforts. This is unacceptable and unjust. There is no excuse for not providing the support now.

A Public Policy-Engaged Social Science

Social scientists have been beating down the policy door for over three decades with the similar rhetoric of needing to include local communities in decision-making processes, having local knowledge guide adaptation, and needing to address the root causes of disasters, climate change, and other adverse events to reduce risk. I repeat this mantra here because while it might appear that such repetition has not changed policy, it has certainly shifted actions and processes in significant ways. For example, international post-disaster recovery processes have moved from focusing solely on valuating physical damages and economic losses to now including impacts on human development.[36] And the 2016 US National Disaster Recovery Framework established coordination across agencies and recognized that local communities have specific cultures, values, and norms that need to be built upon—instead of replaced—by agencies' recovery work.[37] While these might not seem like monumental changes, and the focus is still predominantly on economic systems and the built environment with culture often seen as material and static, and while such frameworks do not guarantee implementation, the policy discourse has certainly shifted, which is a critical start. We need to continue to pound on the policy door.[38] The pathways to change might arise in the most unexpected places and moments. We must be ready.

Participating in a public engaged social science, it is important to keep at the forefront of our minds that because of long standing racial, ethnic, linguistic, and cultural biases against historically and politically marginalized communities, social scientists and other researchers and practitioners can be privileged with policy-related opportunities not often provided to local community members. This privilege should not be taken lightly. It needs to be respected by those holding it as a sense of responsibility and commitment to the communities with whom we engage and work.

Acknowledgment

A special note of gratitude to Chris Brunet for contributing his story to this book and for sharing the stories and beauty of Isle de Jean Charles with me.

Notes

1 Digital story by Chris Brunet, Isle de Jean Charles, audio recorded on July 2 and July 27, 2012.
2 Wong (2016).
3 Ferris (2011).
4 Correa (2011).
5 de Sherbinin et al. (2011).
6 For example, Cernea (1991); de Wet (2006); Cernea and Mathur (2008); Oliver-Smith (2009).
7 Cernea (1997).
8 Cernea (1991, 1997).
9 UNFCCC (2015b).
10 UNFCCC (2015a).
11 Bronen (2011); Bronen and Chapin (2013).
12 Bronen (2011); Bronen and Pollock (2017).
13 Office of the United Nations High Commissioner for Human Rights (2009); Nansen Initiative (2015).
14 Bronen et al. (2018: 259).
15 Hoffman (1999).
16 Fullilove (2005: 14).
17 Maldonado and Peterson (2018).
18 For example, see Peninsula Principles (2013).
19 Stammler (2007).
20 Laska and Peterson (2013).
21 Bronen (2011); Maldonado et al. (2013).
22 Marino and Lazrus (2015).
23 Maldonado (2016b).
24 Maldonado (2016a).
25 Harris and Doerfel (2016); Maldonado (2016b).
26 Maldonado (2016b).
27 Rising Voices (2015).
28 Cash et al. (2002); Maldonado et al. (2016).
29 Maldonado (2016b).
30 Guidelines have been established to foster weaving traditional knowledges and Western science in culturally appropriate, ethical, and respectful ways to address climate change, based on the tenets of free, prior, and informed consent, and cause no harm (CTKW, 2014).
31 Maldonado et al. (2016); Rising Voices (2017).

32 Pulwarty (2013); also Williams and Hardison (2013).
33 Whyte (2013); Williams and Hardison (2013).
34 Powell and Maldonado (2017).
35 Whyte, forthcoming.
36 UN Development Group et al. (2013).
37 US Department of Homeland Security (2016).
38 Maldonado (2016b).

CONCLUSION

No community with a sense of justice, compassion or respect for basic human rights should accept the current pattern of adaptation. . . . As climate change destroys livelihoods, displaces people and undermines entire social and economic systems, no country—however rich or powerful—will be immune to the consequences.

—Archbishop Desmond Tutu[1]

The preceding chapters focused on the lived experience of rapid environmental change engineered through a specific sociopolitical and historical foundation. This context, established at the start of the colonial settler process, has manufactured the vulnerability of coastal Louisiana's communities and transformed the region into an energy sacrifice zone that risks setting the stage for future disasters, and unprecedented changes to our climate system.

The data showed how environmental degradation and state-led coastal restoration and hurricane protection plans reflected and reproduced social inequalities and power dynamics that have turned coastal Louisiana into an energy sacrifice zone. The contemporary rapid environmental change and resulting consequences are a continuation of the structural violence the tribes in coastal Louisiana have faced for generations. It is not about a single hurricane, oil spill, or flood event. Rather, the effects result from a legacy of atrocities and layered processes of systematic discrimination, unsustainable, capitalist-based extractive practices, and industry/government relations around fossil fuel extraction, production, and control that have led to an increasingly changing climate, and put the tribal communities at risk of displacement. Thus, occurrences such as the 2010 BP Deepwater Horizon Oil Disaster and the politically managed cleanup process that further contaminated the environment and affected people's health and livelihoods are not isolated events, but rather are part of a greater, socially constructed disaster.

★ ★ ★

Fighting to Save Home

By Chief Shirell Parfait-Dardar, Grand Caillou/Dulac[2]

You never know the meaning of "Home" until you've lived in more than one. I've been many places in my short lifetime, different States, another Country, different people and surroundings. But I've always longed for and returned "Home."

There's nothing like the smell of the fresh morning air when you first step outside your front door down the bayou. The dampened leaves and grass, and the smell of the bayou right across your street. If you could ever smell peacefulness, that would be it.

I belong to the Grand Caillou/Dulac Band of Biloxi-Chitimacha-Choctaw Indians. I am proud of my heritage and culture. My people are proud and strong. We instill this in our children from birth as we must overcome many obstacles in our bayou homeland of Louisiana. We have to fight for education opportunities for our children, jobs, our land and our way of living.

The loss of our homeland is the hardest battle to overcome, but we will never give up. We are relentless! From land being swindled back in the 1800s to the scariest villain yet, land erosion. We lose football fields every day! The government doesn't do anything to save our land and neither do the oil companies who are responsible for digging the damaging canals many years ago. These companies reap the rewards while we are left to sink into the Gulf of Mexico. They didn't rape Mother Earth; they have destroyed her and my people and our heritage with their lack of morality and common sense. These people have completely missed the concept of my favorite proverb, "We do not own the land. It was not given to us by our ancestors; it was loaned to us by our children."

There were once many trees on our land and they sink away or die from saltwater intrusion. The vastly wooded areas we used to run barefoot in as children are becoming less and less as time passes. Things are changing so quickly I fear that my grandchildren will never know the joys of what I experienced as an Indian child living on the bayou.

They say we can't be saved. They say that we have to relocate to preserve life. Once again they have proven us to be expendable in their eyes. But we have proven that we never back down from the impossible. We will continue to fight for our way of life, for ours is unlike any other. We will fight until the last tree has died, until the last bit of land has washed away. We will always fight for our "Home."

FIGURE 9.1 Chief Shirell, standing where she grew up in Dulac.

Source: Photo by author, 2012.

FIGURE 9.2 Minimal habitat remains for native species in Grand Caillou/Dulac.

Source: Photo by author, 2012.

★ ★ ★

Overcoming Colonial Legacies and the Engineering of a Sacrifice Zone

> It's just going to be sacrificed. It always makes you figure you're just being sacrificed for bigger benefits.
>
> *Gabrielle, Isle de Jean Charles*[3]

The legacy in the United States of privatizing Earth's waters, lands, and minerals for corporate profit have continued to subjugate and marginalize tribal communities like Isle de Jean Charles, Point-au-Chien, and Grand Caillou/Dulac, as well as many other communities. As the resources of the Gulf and its life-giving estuaries continue to be commodified, the prevailing development regime has rendered it acceptable to ravage the environment for corporate profit. With corporations exercising the rights of citizens, one is left to question which citizens our governance system serves and protects. These processes indicate that by destroying the environment—and the possibility of physical and cultural reproduction within that environment—we are also destroying select populations to make way for profit. The continued insatiable consumption, production, and extraction of fossil fuels are acts of purposeful and harmful negligence. The resulting environmental and cultural degradation is a form of tacit persecution.

Parallel relocations from over 200 years ago during the era of Indian Removal are occurring once again, just now in a reverse geographic direction. Whereas before the tribes settled south at the ends of the bayous, some tribal residents are now moving farther inland away from coastal flooding.

Local residents voiced their frustration over the injustice of continued displacement and relocation. Such frustration was apparent as I sat with Chief Shirell under her elevated porch on a rainy spring afternoon. My computer was on my lap. I pressed "return" so we could listen to her digital story recording together. When I heard the lines, "They say we can't be saved. They say that we have to relocate to preserve life" I asked her what she meant. She explained,

> Big industries like the oil companies, government, local government within the parish and state and federal government. This has been going on since the Trail of Tears. Get out of here is what they're saying. I know when our ancestors, Houma Courteau and his family came and established here, they did not think we'd have to run again. They came here and said wow, this place is great. We won't be bothered here. This is finally an area we can flourish in. And unfortunately his children and great-grandchildren are faced with that again. We have to leave. We're being forced out by the damage that the big companies have caused, by the lack of common sense and effort by our government. . . . It's unfortunate that even in the twenty-first century we're still fighting this.[4]

Since the onset of colonization, tribes in coastal Louisiana have accrued layers of habitual, chronic, and co-occurring disasters and injustices. Residents report a

number of social and cultural impacts, including the loss of a shared family livelihood, diminished sharing and resource exchange, and loss of traditional medicines,
traditions and cultural practices, social networks, social cohesion, reciprocity, and
mutual aid.[5] Yet, many residents take action to resist and/or adapt to challenging
circumstances. They elevate their houses, document impacts, lease waters back from
the oil companies so their families can continue to access the water for fishing,
form partnerships and coalitions, and, in extreme circumstances, pursue proactive
community-led resettlement to a safer area still close enough to their culturally
significant resources.[6]

Reflecting on her time with the communities during a fellowship gathering to
discuss common resource concerns, a tribal representative from Minnesota noted,
"Through everything these communities have been through, they're not whiny,
they're looking to solve their problems, they're looking to have people not bail them
out but just to help them, to understand who they are."[7]

It Is Time for a Cultural Shift

As drilling is once again initiated at the site of the 2010 BP Deepwater Horizon Oil Disaster, the problem will clearly not be solved by environmental
policies alone. Government officials who back the interests of private corporations over the rights of local communities must be held accountable and
the fossil fuel companies need to pay a fair share of the costs accrued to communities. Subsistence livelihoods need to be supported, instead of destroyed.
People need access to economic alternatives rather than being forced into a,
capitalist-based marketplace built on the platform of ever-increasing consumption and extraction.

We are already committed to some amount of climate change and resulting
impacts. The associated climate effects from past CO_2 emissions will persist for
decades and even millennia. As NASA scientist James Hansen (2007) testified to
the US House of Representatives, climate change is driven by cumulative—not
current—emissions. Actions taken today will determine the severity of climate
change impacts for coming decades, centuries, and even millennia. We—including
the current and future generations—will face unknown consequences in this
grand human experiment rendered on the Earth's climate system, as tipping
points occur and we cross thresholds that lead to ever increasingly large and
unprecedented impacts.[8]

We continue on this path without fully understanding the depths of the consequences or, where we do have understanding, choosing to ignore what we know.
We cannot solely rely on creating technological fixes in time to innovate our way
out of immeasurable human and economic costs. The future costs of *inaction* now
will be immeasurable.

In the spirit of Desmond Tutu's reflection at the beginning of this chapter, this
is not a developed vs. developing country or global north vs. global south issue.
There are shared experiences among communities around the world that bring

to the forefront the need to address the injustices endured by historical racism, marginalization, and oppression, and the policy changes that are urgently required today.

Capitalist-driven processes are not inevitable; they only exist because they have been so engrained in many of us as being natural that we forget there are other socio-economic structures could take their place.[9] It is time for a cultural shift from the unsustainable ways resources are extracted and used towards a future that ensures the rights of the people and the environments in which we live for generations to come. Our way forward must integrate a transformed social perspective that sees Mother Earth not as an object to be controlled and ravaged in the name of profit, but as a shared home to be respected and cherished. We can start on a new, sustainable path by looking to some of those most affected by such injustices—frontline communities who have dwelled in place for centuries—and respectfully ask how their knowledge, wisdom, strategies, and actions can guide us all. The question remains whether the institutionalized violence will perpetuate or if local and Indigenous knowledges of the lands and waters will be respected and guide efforts to restore the land, adapt to environmental changes, and re-imagine ways to live more sustainably in a world faced with the ever-increasing consequences of a changing climate.

★ ★ ★

Some days I take walks with my older daughter. We don't go anywhere in particular. We just walk. When we get to a tree, she scrambles up, down, and all around the roots, as if searching for a beginning and end. She waves to the leaves as we pass saying, "Good-bye leaves." When we get to a small stream (which is rare in drought conditions), she looks down at the running water and says, "Hello water." I realize how much I have to learn from my child, lessons about how connected we—humans and non-humans—all are.

At the end of the day, we only have this one home. Some of us have used an over-abundance of its gifts, resulting in harm to other inhabitants. And some of us have gained privilege through the dispossession and forced sacrifice of Indigenous and other frontline communities. We have a responsibility to understand where this privilege comes from, its connection to the current climate crisis, and why taking actions cloaked within the capitalist-colonial structure will continue to result in harm.

Recently we have witnessed—all around the world—growing threats to common lands, air, waters, and health. Yet we have also witnessed mass uprisings, determined to protect our most basic, vital, life-giving resources. It is essential to listen to the voices of people who hold key knowledge, wisdom, and expertise on how to resist injustices of all forms, reduce associated risks, and adapt to the changes coming down the pipeline. We need to work together across communities, cultures, geographic boundaries, and bring diverse knowledge systems to the table, respecting and honoring the tensions and differences, to tell a story of what we are for: justice for the present and future—for all generations and relations now and for those to come.

Acknowledgment

A special note of gratitude to Shirell Parfair-Dardar for contributing her story "Fighting to Save Home" to this book and for teaching me what it means to walk quietly but proudly.

Notes

1 Tutu (2007: 166).
2 Digital story, audio recorded on May 9, 2012. Available online at https://www.youtube.com/watch?v=rbphUxHHIDY&t=8s
3 Conversation with Gabrielle in Houma, Louisiana, July 29, 2012.
4 Conversation with Chief Shirell Parfait-Dardar in Chauvin, Louisiana, May 16, 2012.
5 Laska et al. (2010); Laska (2012); Maldonado et al. (2015); Peterson and Maldonado (2016).
6 Maldonado et al. (2015).
7 Conversation with fellowship participant from Minnesota, in Grand Bayou, Louisiana, March 6, 2012.
8 USGCRP (2017).
9 Marx (1994/1888).

APPENDIX

Additional Resources

The following is a list of websites and select information related to the tribal communities in coastal Louisiana:

Biloxi-Chitimacha-Choctaw of Louisiana www.biloxi-chitimacha.com/tribal_communities.htm

Can't Stop the Water is a documentary film and accompanying website about Isle de Jean Charles. www.cantstopthewater.com/

First People's Conservation Council of Louisiana is an association that was formed to provide a forum for four State Recognized Native American Tribes and their respective Tribal Communities located in Coastal Louisiana to identify and solve natural resource issues on their Tribal lands. Through a strong partnership with the USDA-NRCS, the Council reviews and recommends proposals for conservation projects for Louisiana Tribal Members. http://fpcclouisiana.org/

Grand Caillou/Dulac Band of Biloxi-Chitimacha-Choctaw Tribe www.facebook.com/pages/Grand-CaillouDulac-Band-of-Biloxi-Chitimacha-Choctaw-Indians/167836366609938?sk=wall

Isle de Jean Charles Band of Biloxi-Chitimacha-Choctaw Tribe http://isledejeancharles.com/

Pointe-au-Chien Indian Tribe http://pactribe.tripod.com/

The following is a selected list of organizations and networks relevant to the issues addressed in this book:

Climate Adaptation Knowledge Exchange (CAKE) aims to build a shared knowledge base for managing natural and built systems in the face of rapid climate change. Just as important, it is intended to help build an innovative community of practice. www.cakex.org/

Cultural Survival supports a movement of empowered Indigenous Peoples organizing their communities to engage the international processes, national policies and human rights bodies to respect, protect, and fulfill their rights. www.culturalsurvival.org/

Culture and Disaster Action Network (CADAN) brings academics and practitioners together to integrate culturally relevant strategies into the work of disaster risk reduction and recovery. CADAN members are involved in projects demonstrating how cultural considerations can be efficiently integrated into disaster response and recovery efforts to improve outcomes. http://cultureanddisaster.org/

Deep South Center for Environmental Justice (DSCEJ) was founded in collaboration with community environmental groups and universities within the region to address issues of environmental justice. The DSCEJ Community/University Partnership, under the auspices of Dillard University in New Orleans, provides opportunities for communities, scientific researchers, and decision makers to collaborate on programs and projects that promote the rights of all people to be free from environmental harm as it impacts health, jobs, housing, education, and general quality of life. www.dscej.org/index.php

Earthworks is dedicated to protecting communities and the environment from the adverse impacts of mineral and energy development while promoting sustainable solutions. www.earthworksaction.org/

Environmental Justice and Climate Change Initiative is a coalition of leading organizations and voices for Climate Justice. www.ejcc.org/

Gulf Organized Fisheries in Solidarity & Hope (GO FISH) is a coalition of grassroots organizations from across the Gulf Coast that banded together after the Deepwater Horizon Oil Spill to advocate for the rights of fishermen, restore the fisheries, and preserve fishing community culture. https://eng2viet.files.wordpress.com/2011/07/go-fish.pdf

Gulf Restoration Network is committed to uniting and empowering people to protect and restore the natural resources of the Gulf Region. https://healthygulf.org/

Gulf South Rising (GSR) is a coordinated regional movement created to highlight the impact of the global climate crisis on the Gulf South region (Texas, Louisiana, Mississippi, Alabama, and Florida). GSR demands a just transition away from extractive industries, discriminatory policies, and unjust practices that hinder equitable disaster recovery and impede the development of sustainable communities. www.gulfsouthrising.org/

Honor the Earth creates awareness and support for Native environmental issues and to develop needed financial and political resources for the survival of sustainable Native communities. www.honorearth.org/

Indigenous Environmental Network (IEN) is an alliance of Indigenous Peoples whose shared mission is to protect the sacredness of Earth Mother from contamination and exploitation by respecting and adhering to Indigenous Knowledge and natural law. www.ienearth.org/

Indigenous Phenology Network is a grassroots organization whose participants are interested in understanding phenology on lands and species of importance to native peoples. The group is focused on building relationships, ensuring benefit to indigenous communities, and integrating indigenous and western knowledge systems. www.usanpn.org/nn/indigenous-phenology-network

Institute for Tribal Environmental Professionals (ITEP) strengthens tribal capacity and sovereignty in environmental and natural resource management through culturally relevant education, research, partnerships and policy-based services. www7.nau.edu/itep/main/Home/

Livelihoods Knowledge Exchange Network (LiKEN) is a link-tank for policy-relevant research toward post-carbon livelihoods and communities. http://likenknowledge.org/

Lowlander Center, based in the bayous of Louisiana, supports lowland people and places through education, research and advocacy. It is a Center that is based on community participatory principles and methods. The work of the Lowlander Center is to help create solutions to living with an ever-changing coastline and land loss while visioning a future that builds capacity and resilience for place and people. www.lowlandercenter.org/

Nansen Initiative is a state-led consultative process to build consensus on a Protection Agenda addressing the needs of people displaced across borders in the context of disasters and climate change. www.nanseninitiative.org/

Northeast Indigenous Climate Resilience Network, organized by the Sustainable Development Institute at the College of Menominee Nation, seeks to convene Indigenous peoples to identify threats to Indigenous self-determination and ways of life and to formulate adaptation and mitigation strategies, dialogues and educational programs that build Indigenous capacities to address climate-related issues. www.nicrn.org/

Oil Change International is a research, communication, and advocacy organization focused on exposing the true costs of fossil fuels and facilitating the coming transition toward clean energy. http://priceofoil.org/

Pacific Northwest Tribal Climate Change Network fosters communication between tribes, agencies, and other entities about climate change policies, programs, and research needs pertaining to tribes and climate change. The Network meets via conference call on the third Wednesday of each month. http://tribalclimate.uoregon.edu/network/

Rising Voices: Collaborative Science with Indigenous Knowledge for Climate Solutions facilitates cross-cultural approaches for adaptation solutions to extreme weather and climate events, climate variability, and climate change. The program brings together Indigenous and Western scientific professionals, tribal and community leaders, scientific professionals, environmental and communication experts, students, educators, and artists from across the United States and around the world. https://risingvoices.ucar.edu/

Union of Concerned Scientists puts rigorous, independent science to work, combining technical analysis and effective advocacy to create innovative, practical solutions for a healthy, safe, and sustainable future. www.ucsusa.org/

University of New Orleans, Center for Hazards Assessment, Response, and Technology is an applied social science hazards research center at the University of New Orleans that collaborates with Louisiana communities including the City of New Orleans and its surrounding parishes. www.uno.edu/chart/

The following is a selected list of documents, reports, and guidelines relevant to the issues addressed in this book:

Guidelines for Considering Traditional Knowledges in Climate Change Initiatives are intended to be provisional, to increase understanding of the role of and protections for traditional knowledges in climate initiatives, to provide provisional guidance to those engaging in efforts that encompass traditional knowledges, and to increase mutually beneficial and ethical interactions between tribes and non-tribal partners. https://climatetkw.wordpress.com/guidelines/

Intergovernmental Panel on Climate Change (IPCC) is a scientific body under the sponsorship of the United Nations. It reviews and assesses the most recent scientific, technical, and socio-economic information produced worldwide relevant to the understanding of climate change. http://ipcc.ch/

Jemez Principles for Democratic Organizing were outlined and adopted at coalition meeting in 1996 hosted by the Southwest Network for Environmental and Economic Justice (SNEEJ), in Jemez, New Mexico. The meeting was designed to reach a common understanding between participants from different cultures, politics, and organizations. www.ejnet.org/ej/jemez.pdf

National Disaster Recovery Framework is a guide that enables effective recovery support to disaster-impacted states, Tribes, and territorial and local jurisdictions. It provides a flexible structure that enables disaster recovery managers to operate in a unified and collaborative manner. It also focuses on how best to restore, redevelop and revitalize the health, social, economic, natural and environmental fabric of the community and build a more resilient Nation. www.fema.gov/media-library-data/1466014998123-4bec8550930f774269e0c5968b120ba2/National_Disaster_Recovery_Framework2nd.pdf

National Response Framework is a guide to how the Nation responds to all types of disasters and emergencies. www.fema.gov/media-library-data/20130726-1914-25045-1246/final_national_response_framework_20130501.pdf

Peninsula Principles on Climate Displacement Within States take into consideration that individual communities are expected to play a fundamental role in organizing themselves and outlining their future needs in regards to a looming—or ever present—climate displacement threat. The Peninsula Principles acknowledge that these communities must come forward with their claims when it comes to climate displacement, but also outline what the corresponding obligations of States are, based within currently existing Human Rights laws, to protect and respect the rights of those affected by climate displacement. http://displacementsolutions.org/wp-content/uploads/2014/12/Peninsula-Principles.pdf

Post-Disaster Needs Assessments, the European Union, the UN Development Group, and the World Bank have collaborated on the development of guides for conducting Post Disaster Needs Assessments (PDNA) and for preparing Disaster Recovery Frameworks (DRF). Both guides are based on good practices and experiences from around the world, and are intended to coalesce international and local support behind a single, government-led post disaster recovery process. www.undp.org/content/undp/en/home/librarypage/crisis-prevention-and-recovery/pdna.html

Sendai Framework for Disaster Risk Reduction is a 15-year, voluntary, non-binding agreement which recognizes that the State has the primary role to reduce disaster risk but that responsibility should be shared with other stakeholders including local government, the private sector and other stakeholders. It aims for the substantial reduction of disaster risk and losses in lives, livelihoods and health and in the economic, physical, social, cultural and environmental assets of persons, businesses, communities and countries. www.unisdr.org/files/43291_sendaiframeworkfordrren.pdf

United Nations Declaration on the Rights of Indigenous Peoples is the most comprehensive international instrument on the rights of indigenous peoples. It establishes a universal framework of minimum standards for the survival, dignity and well-being of the indigenous peoples of the world and it elaborates on existing human rights standards and fundamental freedoms as they apply to the specific situation of indigenous peoples. www.un.org/esa/socdev/unpfii/documents/DRIPS_en.pdf

United Nations Guiding Principles on Internal Displacement provide an advocacy and monitoring framework for the assistance and protection needs of the internally displaced. www.unhcr.org/43ce1cff2.html

United States National Climate Assessment (NCA) informs the Nation about observed changes, the current status of the climate, and anticipated trends for the future; integrates scientific information from multiple sources and sectors to highlight key findings and significant knowledge gaps; establishes consistent methods for evaluating climate impacts in the United States in the context of broader global change; and is used by the US government, communities, and businesses as they create more sustainable and environmentally sound plans for the future. www.globalchange.gov/what-we-do/assessment

REFERENCES

Adamo, Susan, and Alex de Sherbinin. 2009. *Environmentally Induced Population Displacements*. CARE-PRC Briefing, 11 February, Washington, DC.

Akers, Donna. 1999. Removing the Heart of the Choctaw People: Indian Removal From a Native Perspective. *American Indian Culture and Research Journal* 23(3): 63–76.

Albrecht, Glenn, Gina-Maree Sartore, Linda Connor, Nick Higginbotham, Sonia Freeman, Brian Kelly, Helen Stain, Anne Tonna, and Georgia Pollard. 2007. Solastalgia: The Distress Caused by Environmental Change. *The Royal Australian and New Zealand College of Psychiatrists* 15: S95–S98.

Apache Corporation. 2010. *More Than Meets the Eye: Summary Annual Report*. [Available online at www.apachecorp.com/Resources/Upload/file/downloads/Apache_AR_2010. pdf].

———. 2014a. *History: 1954–1979*. [Available online at www.apachecorp.com/About_ Apache/History/1954-1979.aspx].

———. 2014b. *Investors*. [Available online at www.apachecorp.com/Investors/index.aspx].

Arctic Council. 2009. *Tromsø Declaration*. On the occasion of the Sixth Ministerial Meeting of the Arctic Council, the 29th of April, 2009, Tromsø, Norway. [Available online at www. state.gov/e/oes/rls/other/2009/123483.htm].

Arrhenius, Svante. 1896. On the Influence of Carbonic Acid in the Air Upon the Temperature of the Ground. *The London, Edinburgh and Dublin Philosophical Magazine and Journal of Science*, Series 5, 41: 237–276.

Associated Press. 2011. US Judge: Spill Claims Czar Not Independent of BP. *USA Today*, February 3. [Available online at http://usatoday30.usatoday.com/money/companies/ regulation/2011-02-02-gulf-spillfeinberg_N.htm].

Austin, Diane. 2006. Cultural Exploitation, Land Loss and Hurricanes: A Recipe for Disaster. *American Anthropologist* 108(4): 671–691.

Auyero, Javier, and Débora Alejandra Swistun. 2009. *Flammable: Environmental Suffering in an Argentine Shantytown*. Oxford: Oxford University Press.

Baram, Marcus. 2011. BP Spent $2 Million Lobbying on Offshore Drilling, Spill Liability, Other Regulations in First Quarter of 2011. *Huffington Post*, April 21. [Available online at www. huffingtonpost.com/2011/04/21/bp-lobbying-2011-q1-2-million_n_851842.html].

Barker, Joanne. 2011. *Native Acts: Law, Recognition, and Cultural Authenticity*. Durham, NC: Duke University Press.

Barry, John M. 1997. *Rising Tide: The Great Mississippi Flood of 1927 and How It Changed America*. New York: Simon and Schuster.

Bartrop, Paul R. 2007. Episodes From the Genocide of the Native Americans: A Review Essay. *Genocide Studies and Prevention* 2(2): 183–190.

Basso, Keith H. 1996. *Wisdom Sits in Places: Landscape and Language Among the Western Apache*. Albuquerque: University of New Mexico Press.

Bethel, Matthew B., Lynn F. Brien, Emily J. Danielson, Shirley B. Laska, John P. Troutman, William M. Boshart, Marco J. Giardino, and Maurice A. Phillips. 2011. *Blending Geospatial Technology and Traditional Ecological Knowledge to Enhance Restoration Decision-Support Processes in Coastal Louisiana*. CHART Publications. Paper 23: 555–571. [Available online at http://scholarworks.uno.edu/chart_pubs/23].

BIA [US Bureau of Indian Affairs]. 2008a. *Summary Under the Criteria and Evidence for Amended Proposed Finding Against Federal Acknowledgment of the Biloxi, Chitimacha Confederation of Muskogees, Inc.* Washington, DC: Bureau of Indian Affairs.

———. 2008b. *Summary Under the Criteria and Evidence for Amended Proposed Finding Against Federal Acknowledgment of the Pointe-au-Chien Indian Tribe*. Washington, DC: Bureau of Indian Affairs.

Bisht, Tulsi Charan. 2011. Resettlement in the Tehri Dam Project: An Ethnographic Profile. In *Resettling Displaced People: Policy and Practice in India*. Hari Mohan Mathur, ed., pp. 245–266. New Delhi: Routledge.

Blackburn, Marion. 2012. Return to the Trail of Tears. *Archaeology* 65(2): 53–64.

Blanchard, Paulette. 2015. Our Squirrels will have Elephant Ears: Indigenous Perspectives on Climate Change in the South Central United States. Master's Thesis, University of Oklahoma. [Available online at https://shareok.org/handle/11244/51935].

Blunden, Jessica, and Derek S. Arndt, eds. 2017. State of the Climate in 2016. *Bulletin of the American Meteorological Society* 98(8): Si–S277. doi:10.1175/2017BAMSStateofthe Climate.1.

Bourgois, Philippe, and Nancy Scheper-Hughes. 2004. Comments on an Anthropology of Structural Violence. *Current Anthropology* 45(3): 317–318.

BP [British Petroleum]. 2014. *Early History*. [Available online at www.bp.com/en/global/corporate/about-bp/our-history/history-of-bp/early-history.html].

Brasseaux, Carl A. 1985. A New Acadia: The Acadian Migrations to South Louisiana, 1764–1803. *Acadiensis: Journal of the History of the Atlantic Region* 15(1): 123–132.

———. 1991. *Scattered to the Wind: Dispersal and Wanderings of the Acadians, 1755–1809*. Lafayette: University of Southwestern Louisiana.

Briscoe Center for American History. 2014. *A Guide to the ExxonMobil Historical Collection, 1790–2004: Part 1. The University of Texas at Austin*. [Available online at www.lib.utexas.edu/taro/utcah/00352/cah-00352.html].

Bronen, Robin. 2011. Climate-Induced Community Relocations: Creating an Adaptive Governance Framework Based in Human Rights Doctrine. *New York University Review of Law and Social Change* 35: 356–406.

———. 2014. Choice and Necessity: Relocations in the Arctic and South Pacific. *Forced Migration Review* 45: 17–21.

Bronen, Robin, and F. Stuart Chapin III. 2013. Adaptive Governance and Institutional Strategies for Climate-induced Community Relocations in Alaska. *PNAS* 110(23): 9320–9325.

Bronen, Robin, Julie Maldonado, Elizabeth Marino, and Preston Hardison. 2018. Climate Change and Displacement: Challenges and Needs to Address an Imminent Reality. In

Challenging the Prevailing Displacement and Resettlement Paradigm: Risks, Impoverishment, Legacies, and Solutions. Michael M. Cernea and Julie K. Maldonado, eds., London and New York Routledge Press.

Bronen, Robin, and Denise Pollock. 2017. *Climate Change, Displacement and Community Relocation: Lessons From Alaska.* Alaska Institute for Justice and Norwegian Refugee Council. [Available online at www.nrc.no/globalassets/pdf/reports/nrc-alaska_relocation-screen. pdf].

Buckley, Geoffrey L., and Laura Allen. 2011. Stories About Mountaintop Removal in the Appalachian Coalfields. In *Mountains of Injustice: Social and Environmental Justice in Appalachia.* Michele Morrone, Geoffrey L. Buckley, and Jedidiah Purdy, eds., pp. 161–180. Athens: Ohio University Press.

Burley, David. 2010. *Losing Ground: Identity and Land Loss in Coastal Louisiana.* Jackson: University Press of Mississippi.

Burley, David, Pam Jenkins, Shirley Laska, and Traber Davis. 2007. Place Attachment and Environmental Change in Coastal Louisiana. *Organization and Environment* 20(3): 347–366.

Button, Gregory. 2001. Popular Media Reframing of Man-made Disasters: A Cautionary Tale. In *Catastrophe and Culture: The Anthropology of Disaster.* Anthony Oliver-Smith and Susanna M. Hoffman, eds., pp. 143–158. Santa Fe, NM: School of American Research Press.

Button, Gregory, and Anthony Oliver-Smith. 2008. Disaster, Displacement and Employment: Distortion of Labor Markets During Post-Katrina Reconstruction. In *Capitalizing on Catastrophe: Neoliberal Strategies in Disaster Reconstruction.* Nandini Gunewardena and Mark Schuller, eds., pp. 123–145. Lanham, MD: Altamira Press.

Cash, D., W. C. Clark, F. Alcock, N. M. Dickson, N. Eckley, and J. Jäger. 2002. *Salience, Credibility, Legitimacy and Boundaries: Linking Research, Assessment and Decision Making.* KSG Working Papers Series RWP02–046. [Available online at http://dx.doi.org/10.2139/ssrn.372280].

Castles, Stephen. 2002. *Environmental Change and Forced Migration: Making Sense of the Debate. New Issues in Refugee Research.* Working Paper No. 70. Refugees Studies Centre. University of Oxford, UK.

Center for Biological Diversity. 2014. *Dispersants.* [Available online at www.biological diversity.org/programs/public_lands/energy/dirty_energy_development/oil_and_gas/gulf_oil_spill/dispersants.html].

Cernea, Michael M. 1991. Involuntary Resettlement: Research, Policy, and Planning. In *Putting People First: Sociological Variables of Development.* M. M. Cernea, ed., pp. 188–215. New York: Oxford University Press.

———. 1997. The Risks and Reconstruction Model for Resettling Displaced Populations. *World Development* 25(10): 1569–1587.

Cernea, Michael M., and Hari Mohan Mathur, eds. 2008. *Can Compensation Prevent Impoverishment? Reforming Resettlement Through Investments and Benefit-Sharing.* Oxford: Oxford University Press.

CLEAR [Coastal Louisiana Ecosystem Assessment and Restoration]. 2006. *Reducing Flood Damage in Coastal Louisiana: Communities, Culture and Commerce.* [Available online at www.clear.lsu.edu].

Clement, Joel. 2017. I'm a Scientist. I'm Blowing the Whistle on the Trump Administration. *The Washington Post,* July 19. [Available online at www.washingtonpost.com/opinions/im-a-scientist-the-trump-administration-reassigned-me-for-speaking-up-about-climate-change/2017/07/19/389b8dce-6b12-11e7-9c15-177740635e83_story.html?utm_term=.c814dcd55ee7].

Coastal Louisiana Tribal Communities. 2012. Stories of Change: Coastal Louisiana Tribal Communities' Experiences of a Transforming Environment (Grand Bayou, Grand Caillou/ Dulac, Isle de Jean Charles, Pointe-au-Chien). In *Workshop Report Input Into the National Climate Assessment*, Julie Koppel Maldonado, ed. Pointe-aux-Chenes, LA, January 22–27.

Connor, Linda, Glenn Albrecht, Nick Higginbotham, Sonia Freeman, and Wayne Smith. 2004. Environmental Change and Human Health in Upper Hunter Communities of New South Wales, Australia. *EcoHealth* 1(Suppl. 2): 47–58.

Correa, Elena. 2011. *Preventive Resettlement of Populations at Risk of Disaster Experiences From Latin America*. Washington, DC: World Bank.

Couvillion, Brady A., John A. Barras, Gregory D. Steyer, William Sleavin, Michelle Fischer, Holly Beck, Nadine Trahan, Brad Griffin, and David Heckman. 2011. *Land Area Change in Coastal Louisiana From 1932 to 2010: US Geological Survey Scientific Investigations Map 3164*. [Available online at http://pubs.usgs.gov/sim/3164/downloads/SIM3164_Pamphlet.pdf].

CPRA [Coastal Protection and Restoration Authority of Louisiana]. 2012. *Louisiana's Comprehensive Master Plan For a Sustainable Coast*. Baton Rouge, LA: CPRA. [Available online at www.lacpra.org/assets/docs/2012%20Master%20Plan/Final%20Plan/2012%20Coastal%20Master%20Plan.pdf].

———. 2017. *Louisiana's Comprehensive Master Plan for a Sustainable Coast*. Baton Rouge, LA: CPRA. [Available online at http://coastal.la.gov/wp-content/uploads/2017/04/2017-Coastal-Master-Plan_Web-Single-Page_CFinal-with-Effective-Date-06092017.pdf].

CTKW [Climate and Traditional Knowledges Workgroup]. 2014. *Guidelines for Considering Traditional Knowledges in Climate Change Initiatives*. [Available online at https://climatetkw.wordpress.com].

Cubitt, Geoffrey. 2007. *History and Memory*. Manchester, UK: Manchester University Press.

Cummins, Light Townsend. 2014. Part One. In *Louisiana: A History*. Bennett H. Wall and John C. Rodrigue, eds., pp. 7–104. Sussex, UK: Wiley Blackwell.

Davenport, Coral, and Campbell Robertson. 2016. Resettling the First American 'Climate Refugees'. *The New York Times*, May 3. [Available online at www.nytimes.com/2016/05/03/us/resettling-the-first-american-climate-refugees.html?mcubz=0].

De Melker, Saskia. 2012. *Quileute Tribe Fights to Regain Precious Land in a Changing Climate*. Oregon Public Broadcasting, August 14. [Available online at www.opb.org/news/article/quileute-relocate-climate-change-twilight-video/].

De Sherbinin, A., M. Castro, F. Gemenne, M.M. Cernea, S. Adamo, P.M. Fearnside, G. Krieger, S. Lahmani, A. Oliver-Smith, A. Pankhurst, T. Scudder, B. Singer, Y. Tan, G. Wannier, P. Boncour, C. Ehrhart, G. Hugo, B. Pandey, and G. Shi. 2011. Preparing for Resettlement Associated With Climate Change. *Science* 334: 456–457.

De Wet, Chris, ed. 2006. *Development-induced Displacement: Problems, Policies and People*. New York: Berghahn Books.

Displacement Solutions. 2015. One Step at a Time: The Relocation Process of the Gardi Sugdub Community in Gunayala, Panama. Mission Report. [Available online at http://displacementsolutions.org/wp-content/uploads/2015/07/One-Step-at-a-Time-the-Relocation-Process-of-the-Gardi-Sugdub-Community-In-Gunayala-Panama.pdf].

Displacement Solutions and YPSA/Young Power in Social Action. 2014. *Climate Displacement in Bangladesh: Stakeholders, Laws and Policies—Mapping the Existing Institutional Framework*. Bangladesh Housing, Land and Property (HLP) Rights Initiative. [Available online at http://displacementsolutions.org/wp-content/uploads/Mapping-Study-Climate-Displacement-Bangladesh.pdf].

DOI [US Department of the Interior—Bureau of Indian Affairs]. 2015. *Final Rule for 25 CFR Part 83 Acknowledgment of American Indian Tribes.* [Available online at www.bia.gov/cs/groups/public/documents/text/idc1-030742.pdf].

Donini, Antonio. 2008. Through a Glass, Darkly: Humanitarianism and Empire. In *Capitalizing on Catastrophe: Neoliberal Strategies in Disaster Reconstruction.* N. Gunewardena and M. Schuller, eds., pp. 29–44. Lanham, MD: Alta Mira.

Dowd, Gregory Evans. 2004. The American Revolution to the Mid-Nineteenth Century. In *The Handbook of North American Indians: Volume 14, Southeast.* Raymond D. Fogelson, ed., pp. 139–151. Washington, DC: Smithsonian Institution Press.

D'Souza, Rohan. 2008. Framing India's Hydraulic Crises: The Politics of the Modern Large Dam. *Monthly Review* 60(3): 112–124.

Dupre, Senator Reggie. 2004. *Senate Concurrent Resolution No. 105.* [Available online at www.legis.state.la.us/leg_docs/04RS/CVT6/OUT/0000LPU6.PDF].

EESI [Environmental and Energy Study Institute]. 2015. *What If the Water Can't Be Stopped? Tribal Resilience Plans in an Age of Sea Level Rise.* Congressional Briefing, Washington, DC, April 20. [Available online at www.eesi.org/briefings/view/042015tribal].

Environmental Defense Center. 2011. *Fracking—Federal Law: Loopholes and Exemptions.* Santa Barbara, CA: Environmental Defense Center. [Available online at www.edcnet.org/learn/current_cases/fracking/federal_law_loopholes.html].

EPA [US Environmental Protection Agency]. 2012. *Radium.* [Available online at www.epa.gov/radiation/radionuclides/radium.html#effects].

———. 2013. *Wetlands—Wetland Types.* [Available online at http://water.epa.gov/type/wetlands/types_index.cfm].

Family Search. 2014. *Terrebonne Parish, Louisiana. 1930 and 1960 U.S. Census.* [Available online at https://familysearch.org/learn/wiki/en/Terrebonne_Parish,_Louisiana#Census].

Farmer, Paul. 2003. *Pathologies of Power: Health, Human Rights, and the New War on the Poor.* Berkeley: University of California Press.

———. 2004. An Anthropology of Structural Violence. *Current Anthropology* 45(3): 305–325.

Ferris, Elizabeth. 2011. *Humanitarian Silos: Climate Change-induced Displacement.* Washington, DC: Brookings Institution. [Available online at www.brookings.edu/research/humanitarian-silos-climate-change-induced-displacement/].

———. 2012. *Protection and Planned Relocations in the Context of Climate Change.* Geneva, Switzerland: United Nations High Commission of Refugees, Division of International Protection. [Available online at www.unhcr.org/5024d5169.html].

Fortson, Danny 2008. Stern Warns that Climate Change is far Worse than 2006 Estimate. *The Independent*, 17 April. [Available online at www.independent.co.uk/news/business/news/stern-warns-that-climate-change-is-farworse-than-2006-estimate-810488.html].

Frankson, Rebekah, and Kenneth Kunkel. 2017. *Louisiana, State Summaries 149-LA.* NOAA National Centers for Environmental Information. [Available online at https://statesummaries.ncics.org/la].

Freire, Paulo. 1970. *Pedagogy of the Oppressed.* New York: Seabury Press.

Freudenburg, William R., and Robert Gramling. 2011. *Blowout in the Gulf: The BP Oil Spill Disaster and the Future of Energy in America.* Cambridge, MA: The MIT Press.

Freudenberg, William R., Robert Gramling, Shirley Laska, and Kai T. Erikson. 2009. *Catastrophe in the Making: The Engineering of Katrina and the Disasters of Tomorrow.* Washington, DC: Island Press.

Fullilove, Mindy Thompson. 2005. *Root Shock: How Tearing up City Neighborhoods Hurts American and What We Can Do About It.* New York: One World Press.

Gemenne, Francois. 2010. *Climate-induced Population Displacements in a 4°C+ World.* Paper presented at the International Resettlement Conference: Economics, Social Justice, and

Ethics in Development-Caused Involuntary Migration at the 15th International Metropolis Conference, The Hague, Netherlands.

Gill, James. 1989. We Mustn't Make Big Oil Angry. *The Times-Picayune.* New Orleans, LA. B11.

Gremillion, Kristen J. 2004. Environment. In *The Handbook of North American Indians: Volume 14, Southeast.* Raymond D. Fogelson, ed., pp. 53–67. Washington, DC: Smithsonian Institution Press.

Gulf Coast Claims Facility. 2011. *Modification to Final Rules Governing Payment Options, Eligibility and Substantiation Criteria, and Final Payment Methodology.* [Available online at http://gulfcoastdisaster.com/etc/deepwater/files/docs/METHODOLOGY_8162011.pdf].

Gulf Coast Ecosystem Restoration Council. 2013. *Draft Initial Comprehensive Plan: Restoring the Gulf Coast's Ecosystem and Economy.* [Available online at www.restorethegulf.gov/sites/default/files/Gulf%20Restoration%20Council%20Draft%20Initial%20Comprehensive%20Plan%205.23.15.pdf].

Haas, Edward F. 2014. Part Three. In *Louisiana: A History.* Bennett H. Wall and John C. Rodrigue, eds., pp. 233–324. Sussex, UK: Wiley Blackwell.

Hammer, David. 2011. Most BP Oil Spill Claimants Opt for One-Time "Quick Payment." *The Times-Picayune,* January 26. [Available online at http://www.nola.com/news/gulf-oilspill/index.ssf/2011/01/most_bp_oil_spill_claimants_op.html].

Hanley, Steve. 2017. US Interior Department Suppresses Sea Level Rise Study Conclusion. *Clean Technica,* May 24. [Available online at https://cleantechnica.com/2017/05/24/us-interior-department-suppresses-sea-level-rise-study-conclusion/].

Hansen, James. 1988. *The Greenhouse Effect: Impacts on Current Global Temperature and Regional Heat Waves.* Statement of Dr. James Hansen, Director, NASA Goddard Institute for Space Studies. Presented to United States Senate Committee on Energy and Natural Resources, June 23. [Available online at www.columbia.edu/~jeh1/2007/Testimony_20070319.pdf].

———. 2007. *Political Interference With Government Climate Change Science.* Testimony of James E. Hansen to Committee on Oversight and Government Reform United States House of Representatives, March 19. [Available online at www.columbia.edu/~jeh1/2007/Testimony_20070319.pdf].

Hansen, James, M. Sato, P. Hearty, R. Ruedy, M. Kelley, V. Masson-Delmotte, G. Russell, G. Tselioudis, J. Cao, E. Rignot, I. Velicogna, E. Kandiano, K. von Schuckmann, P. Kharecha, A.N. LeGrande, M. Bauer, and K.-W. Lo. 2015. Ice Melt, Sea Level Rise and Superstorms: Evidence From Paleoclimate Data, Climate Modeling, and Modern Observations that 2°C Global Warming Is Highly Dangerous. *Atmospheric Chemistry and Physics* 15: 20059–20179.

Harris, Jack, and Marya Doerfel. 2016. Interorganizational Resilience: Networked Collaborations in Communities After Superstorm Sandy. In *Social Network Analysis in Disaster Response, Recovery, and Adaptation.* E. Jones and A. J. Faas, eds., pp. 75–92. Oxford, UK: Elsevier.

Harrison, Jill Ann. 2012. *Buoyancy on the Bayou: Shrimpers Face the Rising Tide of Globalization.* Ithaca, NY: Cornell University Press.

Harvey, David. 2003. *The New Imperialism.* Oxford: Oxford University Press.

———. 2005. *A Brief History of Neoliberalism.* Oxford: Oxford University Press.

Hauer, Mathew E., Jason M. Evans, and Deepak R. Mishra. 2016. Millions Projected to Be at Risk from Sea-Level Rise in the Continental United States. *Nature Climate Change* 6(7): 691–695.

Hirsch, Mark. 2009. *Thomas Jefferson: Founding Father of Indian Removal.* American Indian. Summer: 55–58. Washington, DC: Smithsonian Institution.

Hoffman, Susanna. 1999. After Atlas Shrugs: Cultural Change or Persistence After a Disaster. In *The Angry Earth: Disaster in Anthropological Perspective.* Anthony Oliver-Smith and Susanna M. Hoffman, eds., pp. 302–325. New York: Routledge.

Holm Jr., Guerry O., Elaine Evers, and Charles E. Sasser. 2011. *The Nutria in Louisiana: A Current and Historical Perspective. Final Report.* Prepared for The Lake Pontchartrain Basin Foundation. [Available online at www.saveourlake.org/PDF-documents/our-coast/LPBF-LSU-Nutria-FINAL-11-22-11.pdf].

Hoover, Herbert. 1975. *The Chitimacha People.* Phoenix, AZ: Indian Tribal Series.

Houck, Oliver A. 2015. The Reckoning: Oil and Gas Development in the Louisiana Coastal Zone. *Tulane Environmental Law Journal* 28(2): 185–296.

Ingles, Palma, and Heather McIlvaine-Newsad. 2007. Any Port in the Storm: The Effects of Hurricane Katrina on Two Fishing Communities in Louisiana. *NAPA Bulletin* 28: 69–86.

International Organization for Migration [IOM]. 2009. *Migration, Climate Change and Environmental Degradation.* [Available online at www.iom.int/jahia/Jahia/pid/2068].

———. 2017. *Migration, Climate Change and the Environment: A Complex Nexus.* [Available online at www.iom.int/complex-nexus#estimates].

IPCC [Intergovernmental Panel on Climate Change]. 1990. *Climate Change: The IPCC Impacts Assessment.* [Available online at www.ipcc.ch/ipccreports/far/wg_II/ipcc_far_wg_II_full_report.pdf].

———. 2014. *Climate Change 2014: Impacts, Adaptation, and Vulnerability. Part A: Global and Sectoral Aspects.* Contribution of Working Group II to the Fifth Assessment Report of the Intergovernmental Panel on Climate Change. [C.B. Field, V.R. Barros, D.J. Dokken, K.J. Mach, M.D. Mastrandrea, T.E. Bilir, M. Chatterjee, K.L. Ebi, Y.O. Estrada, R.C. Genova, B. Girma, E.S. Kissel, A.N. Levy, S. MacCracken, P.R. Mastrandrea, and L.L. White (eds.)]. Cambridge, UK and New York, NY: Cambridge University Press.

Jackson, Andrew. 1830. *A Century of Lawmaking for a New Nation: U.S. Congressional Documents and Debates, 1774–1875, Register of Debates, 21st Congress, 2nd Session.* Message of the President of the United States, to Both Houses of Congress, at the Commencement of the Second Session of the Twenty-first Congress, December 7, 1830. [Available online at http://memory.loc.gov/].

Jackson, Jason Baird, and Raymond D. Fogelson. 2004. Introduction. In *The Handbook of North American Indians: Volume 14, Southeast.* Raymond D. Fogelson, ed., pp. 1–13. Washington, DC: Smithsonian Institution Press.

Jamail, Dahr. 2012. Gulf Seafood Deformities Alarm Scientists. *Al Jazeera,* April 20. [Available online at www.aljazeera.com/indepth/features/2012/04/201241682318260912.html].

Jones, Swanson, Huddell and Garrison, LLC. 2013. *Petition for Damages and Injunctive Relief.* Filed July 24, 2013, Civil District Court for the Parish of Orleans, State of Louisiana. [Available online at http://jonesswanson.com/wpcontent/uploads/2013/07/Petition-for-Damages-and-Injunctive-Relief-with-Exhibits-AG.pdf].

Juhasz, Antonia. 2011. *Black Tide: The Devastating Impact of the Gulf Oil Spill.* Hoboken, NJ: John Wiley and Sons, Inc.

Kachko, Liza. 2013. *Changing Landscapes: Impacts of Environmental Change on Knowledge and Use of Medicinal and Edible Wild Plants in the Communities of Pointe au Chien and Isle de jean Charles in Southern Louisiana.* MSc Thesis, University of Kent at Canterbury.

Kachko, Liza, Theresa Dardar, and Julie Maldonado. 2015. *Revitalizing Traditional Medicinal Plant Knowledge and Use Pointe au Chien Indian Tribe, Louisiana.* Poster presentation, Rising Voices. 2014 Adaptation to Climate Change and Variability: Bringing Together Science and Indigenous Ways of Knowing to Create Positive Solutions. National Center for Atmospheric Research, Boulder, CO, June 30–July 2.

Kälin, Walter. 2008. *The Climate Change—Displacement Nexus.* [Available online at www.reliefweb.int/rw/rwb.nsf/db900sid/SKAI-7GNQV9?OpenDocument&query=climate%20change%20displacement%20nexus].

Kane, Harnett T. 1944. *Deep Delta Country.* New York: Duell, Sloan, and Pearce.

Karl, Thomas R., Jerry M. Melillo, and Thomas C. Peterson, eds. 2009. *Global Climate Change Impacts in the United States: A State of Knowledge Report From the US Global Change Research Program.* Cambridge: Cambridge University Press.

Keener, Victoria W., John J. Marra, Melissa L. Finucane, Deanna Spooner, Margaret H. Smith. 2012. *Climate Change and Pacific Islands: Indicators and Impacts.* Report for the 2012 Pacific Islands Regional Climate Assessment. National Climate Assessment Regional Technical Input Report Series. Washington, DC: Island Press.

Klein, Naomi. 2007. *The Shock Doctrine: The Rise of Disaster Capitalism.* New York: Picador.

Klopotek, Brian. 2011. *Recognition Odysseys: Indigeneity, Race, and Federal Tribal Recognition Policy in Three Louisiana Indian Communities.* Durham, NC: Duke University Press.

Kolmannskog, Vikram. 2008. *Climate of Displacement, Climate for Protection? Migration Scenarios.* DIIS Brief, Copenhagen Danish Institute for International Studies.

Kurtz, Michael L. 2014. Part Four. In *Louisiana: A History.* Bennett H. Wall and John C. Rodrigue, eds., pp. 325–480. Sussex, UK: Wiley Blackwell.

LaDuke, Winona. 2013. *20th Annual Sheinberg Lecture on "Predator Economics, Human Rights, and Indigenous Peoples."* New York University, 13 November. [Available online at www.law.nyu.edu/news/winona-laduke-delivers-the-20th-annual-sheinberg-lecture].

Lancaster, Roger N. 2006. *State of Panic.* Unpublished version.

Landrieu, Senator Mary. 2011. *Gulf Coast Recovery: An Examination of Claims and Social Services in the Aftermath of the Deepwater Horizon Oil Spill.* Hearing Before the Ad Hoc Subcommittee on Disaster Recovery of the Committee on Homeland Security and Governmental Affairs, United States Senate, January 27. [Available online at www.gpo.gov/fdsys/pkg/CHRG-112shrg66618/html/CHRG-112shrg66618.htm].

Laska, Shirley. 2012. Dimensions of Resiliency: Essential, Exceptional, and Scale. *International Journal of Critical Infrastructure* 6(3): 246–276.

Laska, Shirley, Tony Laska, Bob Gough, Jack Martin, Albert Naquin, and Kristina Peterson. 2014. *Indigenous Roots for Sustainable Futures: Proactive Solutions for a Time of Change.* Proposal for Isle de Jean Charles Relocation Planning.

Laska, Shirley, and Kristina Peterson. 2013. Between Now and Then: Tackling the Conundrum of Climate Change. CHART Publications. Paper 32: 5–8. [Available online at http://scholarworks.uno.edu/chart_pubs/32].

Laska, Shirley, Kristina Peterson, Michelle E. Alcina, Jonathan West, Ashey Volion, Brent Tranchina, and Richard Krajeski. 2010. *Enhancing Gulf of Mexico Coastal Communities' Resiliency through Participatory Community Engagement.* CHART Publications. Paper 21. [Available online at http://scholarworks.uno.edu/chart_pubs/21].

Laska, Shirley, George Wooddell, Ronald Hagelman, Robert Grambling, and Monica Teets Farris. 2005. At Risk: The Human, Community, and Infrastructure Resources of Coastal Louisiana. *Journal of Coastal Research* 44: 90–111.

Lasley, Carrie Beth. 2011. *National Response Center Incident Reports in Louisiana 2009–10.* Presentation at the American Public Health Association Annual Meeting, Washington, DC, November 3. Presentation also given at Beading Fellowship, Port Sulphur, LA, March 2012.

Lazrus, Heather. 2012. Sea Change: Island Communities and Climate Change. *Annual Review of Anthropology* 41: 285–301.

LDWF [Louisiana Department of Wildlife and Fisheries]. 2016. *2000–2016 Non-Confidential Shrimp Landings and Trips by Basin.* Louisiana Department of Wildlife and Fisheries Trip Ticket Program.

LMOGA [Louisiana Mid-Continent Oil and Gas Association]. 2014. *Chris John, President.* [Available online at www.lmoga.com/about-us/staff/chris-john/].

Louisiana State Legislature. 1978. *Acts 1978, No. 728, Art. 450. Public Things.* [Available online at www.legis.la.gov/Legis/Law.aspx?d=110625].

Louisiana State Senate. 2014. *Senator Robert Adley—District 36.* [Available online at http://senate.la.gov/Adley/biography.asp].

Low, Setha M. 1992. Symbolic Ties that Bind. In *Place Attachment.* Irwin Altman and Setha M. Low, eds., pp. 165–184. New York: Plenum Press.

Lynn, Kathy, John Daigle, Jennie Hoffman, Frank Lake, Natalie Michelle, Darren Ranco, Carson Viles, Garrit Voggesser, and Paul Williams. 2013. The Impacts of Climate Change on Tribal Traditional Foods. *Climatic Change* 120(3).

Maldonado, Julie Koppel. 2008. *Putting a Price-Tag on Humanity: Development-Forced Displaced Communities' Fight for More than Just Compensation.* MA thesis, Department of Anthropology, American University.

———. 2014a. A Multiple Knowledge Approach for Adaptation to Environmental Change: Lessons Learned From Coastal Louisiana's Tribal Communities. *Journal of Political Ecology* 21: 61–82.

———. 2014b. Everyday Practices and Symbolic Forms of Resistance: Adapting to Environmental Change in Coastal Louisiana. In *Hazards, Risks, and Disasters in Society: A Cross-Disciplinary Overview.* Andrew Collins, ed. Philadelphia, PA: Elsevier Inc.

———. 2016a. Considering Culture in Disaster Practice. *Annals of Anthropological Practice.* A. J. Faas, ed., Special Issue, 40(1): 52–60.

———. 2016b. The Practical and Policy Relevance of Social Network Analysis for Disaster Response, Recovery and Adaptation. In *Social Network Analysis in Disaster Response, Recovery, and Adaptation.* E. Jones and A. J. Faas, eds., pp. 255–268. Oxford, UK: Elsevier.

———. 2017. Corexit to Forget It: The Transformation of Coastal Louisiana Into an Energy Sacrifice Zone. In *ExtrACTION: Impacts, Engagements and Alternative Futures.* Kirk Jalbert, Anna Willow, David Casagrande, Stephanie Paladino, and Jeanne Simonelli, eds. New York, NY: Routledge.

Maldonado, Julie Koppel, Heather Lazrus, Bob Gough, Shiloh-Kay Bennett, Karletta Chief, Carla Dhillon, Linda Kruger, Jeff Morisette, Stefan Petrovic, and Kyle Powys Whyte. 2016. The Story of Rising Voices: Facilitating Collaboration Between Indigenous and Western Ways of Knowing. In *Responses to Disasters and Climate Change: Understanding Vulnerability and Fostering Resilience.* Michele Companion and Miriam Chaiken, eds. Boca Raton, FL: CRC Press.

Maldonado, Julie Koppel, Albert P. Naquin, Theresa Dardar, Shirell Parfait-Dardar, and Kelly Bagwell. 2015. Above the Rising Tide: Coastal Louisiana's Tribal Communities Apply Local Strategies and Knowledge to Adapt to Rapid Environmental Change. In *Disasters' Impact on Livelihood and Cultural Survival: Losses, Opportunities, and Mitigation.* Michèle Companion, ed., pp. 239–253. Boca Raton, FL: CRC Press.

Maldonado, Julie Koppel, and Kristina Peterson. 2018. A Community-based Model for Resettlement: Lessons From Coastal Louisiana. In *The Routledge Handbook of Environmental Displacement and Migration.* R. McLeman and F. Gemenne, eds. Routledge Press.

Maldonado, Julie Koppel, Christine Shearer, Robin Bronen, Kristina Peterson, and Heather Lazrus. 2013. The Impact of Climate Change on Tribal Communities in the US: Displacement, Relocation, and Human Rights. *Climatic Change* 120(3): 601–614.

Marino, Elizabeth. 2012. The Long History of Environmental Migration: Assessing Vulnerability Construction and Obstacles to Successful Relocation in Shishmaref, Alaska. *Global Environmental Change* 22(2): 374–381.

———. 2015. *Fierce Climate, Sacred Ground. An Ethnography of Climate Change in Shishmaref, Alaska.* Fairbanks, AK: University of Alaska Press.

Marino, Elizabeth, and Heather Lazrus. 2015. Migration or Forced Displacement?: The Complex Choices of Climate Change and Disaster Migrants in Shishmaref, Alaska and Nanumea, Tuvalu. *Human Organization* 74(4): 341–350.

Markey, Edward. 2010. *Letter to the National Commission on the BP Deepwater Horizon Oil Spill and Offshore Drilling*, September 28.

Marshall, Bob. 2013. New Research: Louisiana Coast Faces Highest Rate of Sea-Level Rise Worldwide. *The Lens*, February 21. [Available online at http://thelensnola.org/2013/02/21/new-research-louisiana-coast-faceshighest-rate-of-sea-level-rise-on-the-planet/].

Marx, Karl. 1994[1888]. The Communist Manifesto; Preface to a Contribution to the Critique of Political Economy; Capital, Volume One. In *Selected Writings*. Lawrence H. Simon, ed., pp. 209–300. Indianapolis, IN: Hackett Publishing.

Maskrey, Andrew, and Walter Gillis Peacock. 1997. A Call for Action: The Hemispheric Congress Speaks on Disaster Reduction and Sustainable Development. *Hemisphere* 8(1): 4–8.

Maynard, Nancy G., ed. 2002. *Native Peoples-Native Homelands Climate Change Workshop Report*. US National Assessment on Climate Change. US Global Change Research Program, NASA Goddard Space Flight Center.

———. 2014. *Native Peoples-Native Homelands Climate Change Workshop II Final Report: An Indigenous Response to Climate Change*. Prior Lake, MN, November 18–21, 2009.

McDonald, Michael J., and John Muldowny. 1982. *TVA and the Dispossessed*. Knoxville: The University of Tennessee Press.

McNamara, Karen E., and Helene Jacot Des Combes. 2015. Planning for Community Relocations Due to Climate Change in Fiji. *International Journal of Disaster Risk Science* 6: 315–319.

McNeil, Bryan T. 2011. *Combating Mountaintop Removal: New Directions in the Fight Against Big Coal*. Urbana: University of Illinois Press.

Melillo, Jerry M., Terese (T.C.) Richmond, and Gary W. Yohe, eds. 2014. *Climate Change Impacts in the United States: The Third National Climate Assessment*. Washington, DC: US Global Change Research Program.

Miller, Mark. 2004. *Forgotten Tribes: Unrecognized Indians and the Federal Acknowledgment Process*. Lincoln: University of Nebraska Press.

Mine, Sarah, Rui Chen, Shelby Shelton, and Marcy Lowe. 2016. *Louisiana Shrimp Value Chain: Price Dynamics, Challenges, and Opportunities*. Datu Research, prepared for the Coalition to Restore Coastal Louisiana. [Available online at www.crcl.org/images/Shrimp.pdf].

Moldenke, Kelsey. 2017. *Taholah Village Relocation Master Plan*. Presentation at the National Adaptation Forum, Minneapolis, MN, 9 May.

Moncrieffe, Joy, and Rosalind Eyben, eds. 2004. *The Power of Labelling: How People Are Categorized and Why It Matters*. London: Earthscan.

Morgan, David W., Nancy I. M. Morgan, and Brenda Barrett. 2006. Finding a Place for the Commonplace: Hurricane Katrina, Communities, and Preservation Law. *American Anthropologist* 108(4): 706–718.

Morris, Christopher. 2012. *The Big Muddy: An Environmental History of the Mississippi and Its Peoples From Hernando De Soto to Hurricane Katrina*. New York: Oxford University Press.

Morton, Robert A., Julie C. Bernier, and John A. Barras. 2006. Evidence of Regional Subsidence and Associated Interior Wetland Loss Induced by Hydrocarbon Production, Gulf Coast Region, USA. *Environmental Geology* 50(2): 261–274.

Morton, Robert A., Julie C. Bernier, John A. Barras, and Nicholas F. Ferina. 2005. *Rapid Subsidence and Historical Wetland Loss in the Mississippi Delta Plain: Likely Causes and Future Implications*. Open-File Report 2005–1216. Reston, VA: US Geological Survey.

Moskowitz, Peter. 2014. As Louisiana Coast Disappears, So Does Landowners' Money. *Aljazeera America*, April 3. [Available online at http://america.aljazeera.com/articles/2014/4/3/louisiana-oil-rights.html].

Mulvey, Kathy, Seth Shulman Dave Anderson, Nancy Cole, Jayne Piepenburg, and Jean Sideris. 2015. *The Climate Deception Dossiers: Internal Fossil Fuel Industry Memos Reveal Decades of Corporate Disinformation.* Union of Concerned Scientists. [Available online at www.ucsusa.org/sites/default/files/attach/2015/07/The-Climate-Deception-Dossiers.pdf].

Nalco. 2014. *Oil and Gas Production and Pipelines.* [Available online at www.nalco.com/eu/industries/oil-gas-production-pipelines.htm].

Nansen Initiative. 2015. *Agenda for the Protection of Cross-Border Displaced Persons in the Context of Disasters and Climate Change, Volume 1.* [Available online at https://nanseninitiative.org/wp-content/uploads/2015/02/PROTECTION-AGENDA-VOLUME-1.pdf].

NAS [National Academy of Sciences]. 2013. Assessing Impacts of the Deepwater Horizon Oil Spill in the Gulf of Mexico. *Science Daily*, July 10. [Available online at www.sciencedaily.com/releases/2013/07/130710122004.htm].

National Commission on the BP Deepwater Horizon Oil Spill and Offshore Drilling. 2011. *Deep Water: The Gulf Oil Disaster and the Future of Offshore Drilling.* Report to the President. [Available online at www.gpo.gov/fdsys/pkg/GPO-OILCOMMISSION/pdf/GPOOILCOMMISSION.pdf].

Native Voices. 2015. *1838: Cherokee Die on Trail of Tears.* [Available online at www.nlm.nih.gov/nativevoices/timeline/296.html].

Newcomb, Steve. 1992. *Five Hundred Years of Injustice: The Legacy of Fifteenth Century Religious Prejudice.* [Available online at www.wmktradio.com/files/Five%20Hundred%20Years%20of%20Injustice.pdf].

NOAA [US National Oceanic and Atmospheric Administration]. 2010. *NOAA and FDA Announce Chemical Test for Dispersant in Gulf Seafood; All Samples Test Within Safety Threshold*, October 29. [Available online at www.noaanews.noaa.gov/stories2010/20101029_seafood.html].

———. 2012. *Global Sea Level Rise Scenarios for the United States National Climate Assessment.* NOAA Technical Report OAR CPO-1. [Available online at http://scenarios.globalchange.gov/sites/default/files/NOAA_SLR_r3_0.pdf].

———. 2013. *Underwater: Land Loss in Coastal Louisiana Since 1932.* [Available online at www.climate.gov/news-features/featured-images/underwater-land-loss-coastallouisiana-1932].

Nonini, Donald M. 2007. *The Global Idea of "The Commons."* New York, NY: Berghahn Books.

NRC [Norwegian Refugee Council]. 2016. *Disasters and Climate Change.* [Available online at www.nrc.no/what-we-do/speaking-up-for-rights/climate-change/].

Office of the United Nations High Commissioner for Human Rights. 2009. *Report of the Office of the United Nations High Commissioner for Human Rights on the Relationship Between Climate Change and Human Rights.* [Available online at http://daccess-ddsny.un.org/doc/UNDOC/GEN/G09/103/44/PDF/G0910344.pdf?OpenElement].

Oliver-Smith, Anthony. 1999. Peru's Five-Hundred-Year Earthquake: Vulnerability in Historical Context. In *The Angry Earth: Disaster in Anthropological Perspective.* Anthony Oliver-Smith and Susanna M. Hoffman, eds., pp. 74–88. New York: Routledge.

———. 2009. *Development and Dispossession: The Crisis of Forced Displacement and Resettlement.* Santa Fe, NM: School for Advanced Research Press.

———. 2010. *Defying Displacement: Grassroots Resistance and the Critique of Development.* Austin: University of Texas Press.

Ong, Aihwa. 1996. Cultural Citizenship as Subject-Making: Immigrants Negotiate Racial and Cultural Boundaries in the United States. *Cultural Anthropology* 37(5): 737–762.

Osborn, Tim. 2013. *Keynote Comments: Critical Needs for Community Resilience.* Presentation at the Building Resilience Workshop IV: Adapting to Uncertainty Implementing Resilience in Times of Change, New Orleans, LA, March 2–9.

Ostrom, Eleanor, Joanna Burger, Christopher B. Field, Richard B. Norgaard, and David Policansky. 1999. Revisiting the Commons: Local Lessons, Global Challenges. *Science* 284: 278–282.

Papiez, Chelsea. 2009. *Climate Change Implications for the Quileute and Hoh Tribes of Washington: A Multidisciplinary Approach to Assessing Climatic Disruptions to Coastal Indigenous Communities.* Master's Thesis, Environmental Studies, The Evergreen State College. [Available online at http://academic.evergreen.edu/g/grossmaz/Papiez_MES_Thesis. pdf].

Park, Peter. 1997. Participatory Research, Democracy, and Community. *Practicing Anthropology* 19(3): 8–13.

Pelletier, Gérard J. 1972. *Ile Jean Charles.* New Orleans, LA.

Peninsula Principles. 2013. *The Peninsula Principles on Climate Displacement Within States.* Displacement Solutions. [Available online at http://displacementsolutions.org/wp-content/uploads/2014/12/Peninsula-Principles.pdf].

Penland, Shea, Lynda Wayne, L.D. Britsch, S. Jeffress Williams, Andrew D. Beall, and Victoria Caridas Butterwortk. 2000. *Process Classification of Coastal Land Loss Between 1932 and 1990 in the Mississippi River Delta Plain, Southeastern Louisiana.* Reston, VA: US Geological Survey.

Perdue, Theda. 2012. The Legacy of Indian Removal. *The Journal of Southern History* 78(1): 3–36.

Peterson, Kristina J. 2011. *Transforming Researchers and Practitioners: The Unanticipated Consequences (Significance) of Participatory Action Research (PAR).* PhD dissertation, Department of Urban Studies, University of New Orleans.

———. 2014. *Community Conversation With Point au Chien Tribe and the OGHS Staff From Presbyterian Church (USA), Tribal Center,* October 28.

Peterson, Kristina J., and Julie K. Maldonado. 2016. When Adaptation Is Not Enough: Between Now and Then of Community-Led Resettlement. In *Anthropology and Climate Change,* 2nd ed. Susan Crate and Mark Nuttall, eds., pp. 336–353. New York, NY: Routledge.

Pink, Sarah. 2009. *Doing Sensory Ethnography.* London: Sage Publications.

Pittman, Craig. 2013. Three Years After BP Oil Spill, USF Research Finds Massive Die-Off. *Tampa Bay Times,* April 4. [Available online at www.tampabay.com/news/environment/water/gulf-oil-spillkilled-millions-of-microscopic-creatures-at-base-of-food/2113157].

Policy Jury Association of Louisiana. 2014. *Parish Government Structure—The Forms of Parish Government.* [Available online at www.lpgov.org/PageDisplay.asp?p1=3010].

Powell, Dana, and Julie Maldonado, eds. 2017. *Just Environmental and Climate Pathways: Knowledge Exchange Among Community Organizers, Scholar-Activists, Citizen-Scientists and Artists.* Society for Applied Anthropology Annual Meeting, Santa Fe, NM, March 28, 2017. [Available online at http://likenknowledge.org/wp-content/uploads/2018/02/Climate-Pathways-Workshop-Report_Santa-Fe_March-2017.pdf].

President's State, Local and Tribal Leaders Task Force on Climate Preparedness and Resilience. 2014. *Recommendations to the President.* Washington, DC. [Available online at www.whitehouse.gov/sites/default/files/docs/task_force_report_0.pdf].

Pulwarty, Roger. 2013. *Presentation at the Rising Voices of Indigenous People in Weather and Climate Science Workshop.* The Indigenous Peoples Climate Change Working Group and National Center for Atmospheric Research, Boulder, CO, July 1–2.

Quileute Nation. 2011. Key Committee Approves Cantwell Bill to Move Quileute Tribe Out of Tsunami Zone. *The Talking Raven: A Quileute Newsletter* 5(16) [Available online at www.quileutenation.org/newsletter/august_2011.pdf].

Quinault Indian Nation. 2015. *Taholah Village Relocation Master Plan.* [Available online at www.quinaultindiannation.com/planning/projectinfo.html].

Quinlan, Paul. 2010. Less Toxic Dispersants Lose Out in BP Oil Spill Cleanup. *The New York Times,* May 13. [Available online at www.nytimes.com/2010/05/13/business/energy-environment/13greenwire-lesstoxic-dispersants-lose-out-in-bp-oil-spil-81183.html?_r=0].

Redsteer, Margaret H., Klara B. Kelley, Harris Francis, and Debra Block. 2010. Disaster Risk Assessment Case Study: Recent Drought on the Navajo Nation, Southwestern United States. In *Annexes and Papers for the 2011 Global Assessment Report on Disaster Risk Reduction.* United Nations. [Available online at www.preventionweb.net/english/hyogo/gar/2011/en/what/drought.html].

Redsteer, Margaret Hiza, Igor Krupnik, and Julie Koppel Maldonado. Forthcoming. Native American Communities and Climate Change. In *Handbook of North American Indians.* I. Krupnik, ed. Washington, DC: Smithsonian Institution Scholarly Press.

Reed Jr., Adolph. 2008. Class Inequality, Liberal Bad Faith, and Neoliberalism: The True Disaster of Katrina. In *Capitalizing on Catastrophe: Neoliberal Strategies in Disaster Reconstruction.* Nandini Gunewardena and Mark Schuller, eds., pp. 147–154. Lanham, MD: Altamira Press.

Reid, Herbert, and Betsy Taylor. 2010. *Recovering the Commons: Democracy, Place, and Global Justice.* Urbana: University of Illinois Press.

Rico-Martínez, Roberto, Terry W. Snell, and Tonya L. Shearer. 2013. Synergistic Toxicity of Macondo Crude Oil and Dispersant Corexit 9500A® to the Brachionus Plicatilis Species Complex (Rotifera). *Environmental Pollution* 173: 5–10.

Rising Voices. 2013. *The Rising Voices of Indigenous People in Weather and Climate Science Workshop.* Indigenous Peoples Climate Change Working Group and National Center for Atmospheric Research, Boulder, CO, July 1–2.

———. 2015. *Rising Voices: Collaborative Science with Indigenous Knowledge for Climate Solutions.* Third Rising Voices Workshop on Learning and Doing: Education and Adaptation Through Diverse Ways of Knowing. National Center for Atmospheric Research, Boulder, CO, June 29–July 1. Workshop report. [Available online at https://risingvoices.ucar.edu/sites/default/files/rv3_report_final.pdf].

———. 2017. *Rising Voices: Collaborative Science with Indigenous Knowledge for Climate Solutions.* Rising Voices 5: Pathways From Science to Action. National Center for Atmospheric Research, Boulder, CO, April 13–15. Workshop report. [Available online at https://risingvoices.ucar.edu/sites/default/files/2017_Rising_Voices5_Report_final.pdf].

Rocheleau, Matt. 2010. As Oil Firms Grow, Response May Slow to Crises Like Gulf Oil Spill. *Christian Science Monitor,* June 18.

Roth, George. 2008. Recognition. In *Handbook of North American Indians. Volume 2: Indians in Contemporary Society.* Garrick A. Bailey, ed., William C. Sturtevant, general editor, pp. 113–128. Washington, DC: Smithsonian Institution.

Rotkin-Ellman, Miriam, Karen K. Wong, and Gina M. Solomon. 2012. Seafood Contamination After the BP Gulf Oil Spill and Risks to Vulnerable Populations: A Critique of the FDA Risk Assessment. *Environmental Health Perspectives* 120(2): 157–161.

Roy, Arundhati. 1999. *The Cost of Living.* New York: The Modern Library.

Sammarco, Paul W., Steve R. Kolian, Richard A. F. Warby, Jennifer L. Bouldin, Wilma A. Subra, and Scott A. Porter. 2013. Distribution and Concentrations of Petroleum Hydrocarbons Associated With the BP/Deepwater Horizon Oil Spill, Gulf of Mexico. *Marine Pollution Bulletin* 73: 129–143.

Saunt, Claudio. 2004. History Until 1776. In *The Handbook of North American Indians: Volume 14, Southeast.* Raymond D. Fogelson, ed., pp. 128–138. Washington, DC: Smithsonian Institution Press.

Schafer, Judith Kelleher. 2014. Part Two. In *Louisiana: A History*. Bennett H. Wall and John C. Rodrigue, eds., pp. 105–232. Sussex, UK: Wiley Blackwell.

Schleifstein, Mark. 2013a. East Bank Levee Authority Votes to Reaffirm Wetlands Damage Lawsuit Against Energy Companies. *The Times-Picayune*, December 5. [Available online at www.nola.com/environment/index.ssf/2013/12/east_bank_levee_authority_vote.html].

———. 2013b. Jindal Opposes Coastal Erosion Lawsuit Due to Oil Industry Contributions, Environmental Groups Say. *The Times-Picayune*, August 28. [Available online at www.nola.com/environment/index.ssf/2013/08/environmental_groups_say_jinda.html].

———. 2017. Appeals Court Rules for Oil Firms, Against Levee Authority in Wetlands Damage Suit. *The Times-Picayune*, March 17. [Available online at www.nola.com/environment/index.ssf/2017/03/dismissal_of_levee_authority_w.html].

Schuller, Mark. 2016. The Tremors Felt Round the World: Haiti's Earthquake as Global Imagined Community. In *Contextualizing Disaster*. Gregory V. Button and Mark Schuller, eds., pp. 66–88. New York: Berghahn Books.

Schuller, Mark, and Julie Maldonado. 2016. Disaster Capitalism. *Annals of Anthropological Practice*. A. J. Faas, ed., Special Issue, 40(1): 61–72.

Schweitzer, Peter, and Elizabeth Marino. 2005. *Shishmaref Co-Location Cultural Impact Assessment*. Seattle: Tetra Tech Inc.

Scudder, Thayer. 2005. *The Future of Large Dams: Dealing With Social, Environmental, Institutional, and Political Costs*. London: Earthscan.

Sheppard, Kate. 2010. BP's Bad Breakup: How Toxic Is Corexit? *Mother Jones*, September/October. [Available online at www.motherjones.com/environment/2010/08/bp-ocean-dispersant-corexit/].

Silverstein, Ken. 2013. Dirty South: The Foul Legacy of Louisiana Oil. *Harper's Magazine*, November: 45–56.

Singer, Merrill. 2011. Down Cancer Alley: The Lived Experience of Health and Environmental Suffering in Louisiana's Chemical Corridor. *Medical Anthropology Quarterly* 25(2): 141–163.

Singer, Merrill, and Scott Clair. 2003. Syndemics and Public Health: Reconceptualizing Disease in Bio-Social Context. *Medical Anthropology Quarterly* 17(4): 423–441.

Smith, Linda Tuhiwai. 2004. *Decolonizing Methodologies: Research and Indigenous Peoples*. London: Zed Books Ltd.

Solet, Kimberly. 2006. *Thirty Years of Change: How Subdivisions on Stilts Have Altered a Southeast Louisiana Parish's Coast, Landscape and People*. Master's thesis, Department of Urban Studies, University of New Orleans.

Solomon, Gina, and Sarah Janssen. 2010. Health Effects of the Gulf Oil Spill. *Journal of the American Medical Association* 304(10): 1118–1119.

Stammler, Florian. 2007. *Relocation Histories and Attachment to Place in Russia's Gas Capital: Field Impressions*. [Available online at www.alaska.edu/move/result/innocom/].

Stein, Sam. 2012. BP's Influence Peddling in Congress Bears Fruit Two Years After Gulf Spill. *Huffington Post*, March 12. [Available online at www.huffingtonpost.com/2012/03/12/bp-oil-spill-gulf-of-mexico-oil-lobbyists_n_1335556.html].

Steiner, Rick. 2015. Letter: Broken Hearts—NOAA Study Is Last Nail in Coffin of Oil Spill Impact. *The Cordova Times*, September 8.

Stern, Nicholas. 2006. *Stern Review: The Economics of Climate Change*. Cambridge, UK: Cambridge University Press.

Stockman, Lorne. 2016. *A Bridge Too Far: How Appalachian Basin Gas Pipeline Expansion Will Undermine US Climate Goals*. Washington, DC: Oil Change International. [Available online at http://priceofoil.org/content/uploads/2016/08/bridge_too_far_report_v6.3.pdf].

Stoffle, Richard W., and Michael J. Evans. 1990. Holistic Conservation and Cultural Triage: American Indian Perspectives on Cultural Resources. *Human Organization* 49(2): 91–99.

Stonich, Susan C., and Peter Vandergeest. 2001. Violence, Environment, and Industrial Shrimp Farming. In *Violent Environments*. Nancy Peluso and Michael Watts, eds., pp. 261–286. Ithaca, NY: Cornell University Press.

Streater, Steve. 2017. BLM 'Priority' List Pushes Drilling, Wall—Leaked Docs. *E&E News-Greenwire*, April 10. [Available online at www.eenews.net/stories/1060052879].

Strom, Mark S., and Rohinee N. Paranjpye. 2000. Epidemiology and Pathogenesis of Vibrio Vulnificus. *Microbes and Infection* 2: 177–188.

Subra, Wilma. 2010. *Testimony of Wilma Subra Before the Subcommittee on Oversight and Investigations of the House Energy and Commerce Committee on Local Impact of the Deepwater Horizon Oil Spill Human Health and Environmental Impacts Associated With the Deepwater Horizon Crude Oil Spill Disaster,* June 7. [Available online at http://democrats.energycommerce. house.gov/sites/default/files/documents/Testimony-Subra-OI-Impact-Deepwater-Oil-Spill-2010–6–7.pdf].

Sunjic, Melia. 2008. *Top UNHCR Official Warns About Displacement from Climate Change,* December 9. [Available online at http://www.unhcr.org/en-us/news/latest/2008/12/493e9bd94/top-unhcr-official-warns-displacement-climate-change.html].

Tao, Zhen, Stephen Bullard, and Covadonga Arias. 2011. High Numbers of Vibrio Vulnificus in Tar Balls Collected From Oiled Areas of the North-Central Gulf of Mexico Following the 2010 BP Deepwater Horizon Oil Spill. *Ecohealth* 8: 507–511.

Taylor, Betsy, Shirley Laska, Kristina Peterson, and Richard Krajeski. 2014. *The Cascading Effects of Disasters on Communities.* Presentation at the Society for Applied Anthropology Meeting, Albuquerque, NM, March 18–22.

Tennessee Valley Authority. 1961. *Floods and Flood Control.* Technical Report No. 26. Knoxville, TN. [Available online at https://ia801702.us.archive.org/22/items/floodsfloodcontr00tenn/floodsfloodcontr00tenn.pdf].

Terrebonne Genealogical Society. 1998. *The First Registered Land Owners of Bayou Little Caillou (Terrebonne Parish, LA).* Terrebonne Life Lines, Vol. 18.

Thornton, Thomas F. 2008. *Being and Place Among the Tlingit.* Seattle: University of Washington Press.

Tom, Stanley. 2012. *Presentation at the First Stewards Symposium.* National Museum of the American Indian, Washington, DC, July 17–20.

Tong, H.E. President Anote. 2014. *Statement.* 69th General Assembly of the United Nations. New York, September 26. [Available online at www.un.org/en/ga/69/meetings/gadebate/pdf/KI_en.pdf].

Truehill, Eric. 1978. Indians Terrebonne's First Settlers. The Houma Daily Courier and the Terrebonne Press, March 5: 2-C.

Turner, R. E. 1997. Wetland Loss in the Northern Gulf Of Mexico: Multiple Working Hypotheses. *Estuaries* 20(1): 1–13.

Tutu, Desmond. 2007/2008. *Fighting Climate Change: Human Solidarity in a Divided World.* Human Development Report 2007/2008, United Nations Development Programme. [Available online at http://hdr.undp.org/sites/default/files/reports/268/hdr_20072008_en_complete. pdf].

UNFCCC [United Nations Framework Convention on Climate Change]. 2015a. *Draft Text on COP 21 Agenda Item 4 (b) Durban Platform for Enhanced Action (decision 1/CP.17) Adoption of a Protocol, Another Legal Instrument, or an Agreed Outcome With Legal Force Under the Convention Applicable to All Parties.* [Available online at https://unfccc.int/resource/docs/2015/cop21/eng/da01.pdf].

————. 2015b. *Paris Agreement.* [Available online at https://unfccc.int/files/essential_background/convention/application/pdf/english_paris_agreement.pdf].

United Nations. 2008. *United Nations Declaration on the Rights of Indigenous Peoples.* [Available online at www.un.org/esa/socdev/unpfii/documents/DRIPS_en.pdf].

United Nations Development Group, World Bank, and European Union. 2013. *Post-Disaster Needs Assessments: Guidelines, Volume A.* [Available online at www.undp.org/content/dam/undp/library/Environment%20and%20Energy/Climate%20Strategies/PDNA%20Volume%20A%20FINAL%2012th%20Review_March%202015.pdf].

USACE [United States Army Corps of Engineers], Louisiana Coastal Protection and Restoration Authority Board, and Terrebonne Levee and Conservation District. 2013. *Final Revised Programmatic Environmental Impact Statement.* Morganza to the Gulf of Mexico, Louisiana. [Available online at www.mvn.usace.army.mil/Portals/56/docs/PD/Projects/MTG/FinalRevisedProgrammaticEISMtoG.pdf].

US Census Bureau. 2000. *Community Facts: Dulac CDP, Louisiana.*

————. 2010a. *American Indians and Alaska Natives in the United States Map.* [Available online at www.census.gov/geo/maps-data/maps/aian_wall_maps.html].

————. 2010b. *Community Facts: Dulac CDP, Louisiana.* [Available online at http://factfinder2.census.gov/faces/nav/jsf/pages/community_facts.xhtml].

US Coast Guard. 2011. *BP Deepwater Horizon Oil Spill: Incident Specific Preparedness Review.* [Available online at www.uscg.mil/foia/docs/DWH/BPDWH.pdf].

US Congress. 1830. *A Century of Lawmaking for a New Nation: US Congressional Documents and Debates, 1774–1875, Statutes at Large, 21st Congress, 1st Session.* Twenty-First Congress. Session I. Chapter 148. Statute I, May 28. [Available online at http://memory.loc.gov/].

US Congress Bicameral Task Force on Climate Change. 2013. *Implementing the President's Climate Action Plan: U.S. Department of Energy Actions the Department of Energy Should Take to Address Climate Change,* August 6. [Available online at www.whitehouse.senate.gov/imo/media/doc/2013-08-06-Bicameral-Task-Force-DOE-Climate-Report.pdf].

US Department of Homeland Security. 2016. *National Disaster Recovery Framework.* [Available online at www.fema.gov/media-library-data/1466014998123-4bec8550930f774269e0c5968b120ba2/National_Disaster_Recovery_Framework2nd.pdf].

US Department of the Interior. 1994. *The Impact of Federal Programs on Wetlands: Volume II.* A Report to Congress by the Secretary of the Interior. [Available online at www.doi.gov/pmb/oepc/wetlands2/v2ch8.cfm].

US Department of the Treasury. 2012. *Resources and Ecosystems Sustainability, Tourist Opportunities, and Revived Economies of the Gulf Coast States Act of 2012.* [Available online at www.treasury.gov/services/restore-act/Documents/Final-Restore-Act.pdf].

US Energy Information Administration. 2016. *Annual Energy Outlook 2016 Early Release: Annotated Summary of Two Cases.* [Available online at www.eia.gov/outlooks/aeo/er/pdf/0383er(2016).pdf].

US Geological Survey. 2008. *Circum-Arctic Resource Appraisal: Estimates of Undiscovered Oil and Gas North of the Arctic Circle.* USGS Fact Sheet 2008–3049, 2008. [Available online at https://pubs.usgs.gov/fs/2008/3049/fs2008-3049.pdf].

USGCRP [US Global Change Research Program]. 2017. *Climate Science Special Report: Fourth National Climate Assessment, Volume I* [Wuebbles, D.J., D.W. Fahey, K.A. Hibbard, D.J. Dokken, B.C. Stewart, and T.K. Maycock (eds.)]. U.S. Global Change Research Program, Washington, DC. [Available online at https://science2017.globalchange.gov/downloads/CSSR2017_FullReport.pdf].

Vélez-Ibáñez, Carlos G. 2004. Regions of Refuge in the United States: Issues, Problems, and Concerns for the Future of Mexican-Origin Populations in the United States. *Human Organization* 63(1): 1–20.

Viosca Jr., Percy. 1928. *Louisiana Wet Lands and the Value of Their Wild Life and Fishery Resources.* Technical Paper No. 6. Department of Conservation, State of Louisiana.

Watts, Michael. 2004. Antinomies of Community: Some Thoughts on Geography, Resources and Empire. *Transactions of the Institute of British Geographers* 29(2): 195–216.

———. 2012. A Tale of Two Gulfs: Life, Death, and Dispossession Along Two Oil Frontiers. *American Quarterly* 64(3): 437–467.

Waxman, Henry, and Bart Stupak. 2010. *Letter to Representative Kathy Castor*, September 1. [Available online at http://democrats.energycommerce.house.gov/sites/default/files/documents/Castor-BPAdvertising-2010-9-1.pdf].

Weber, Harry R. 2011. BP Increases Pay For Claims Czar Ken Feinberg's Law Firm To $1.25 Million Per Month. *Huffington Post*, March 25. [Available online at www.huffingtonpost.com/2011/03/25/bp-kenfeinberg-claims-salary-pay_n_840871.html].

Westerman, Audrey. 2002. *Early Settlers Along Bayou Terrebonne Below Houma.* [Available online at www.rootsweb.ancestry.com/~laterreb/settlers.htm].

Westman, Clinton N. 2013. Social Impact Assessment and the Anthropology of the Future in Canada's Tar Sands. *Human Organization* 72(2): 111–120.

White House. 2010. *Executive Order 13554—Gulf Coast Ecosystem Restoration Task Force*, October 5. [Available online at www.whitehouse.gov/the-press-office/2010/10/05/executive-order-13554-gulf-coast-ecosystem-restoration-task-force].

Whyte, Kyle Powys. 2013. Justice Forward: Tribes, Climate Adaptation and Responsibility in Indian Country. *Climatic Change* 120(3): 517–530.

———. 2014. A Concern About Shifting Interactions Between Indigenous and Non-Indigenous Parties in US Climate Adaptation Contexts. *Interdisciplinary Environmental Review* 15 (2–3): 114–133.

Whyte, Kyle Powys. Forthcoming. Way Beyond the Lifeboat: An Indigenous Allegory of Climate Justice. In *Climate Futures: Reimagining Global Climate Justice.* Debashish Munshi, Kum-Kum Bhavnani, John Foran, and Priya Kurian, eds. University of California Press.

Wildcat, Daniel R. 2009. *Red Alert!: Saving the Planet With Indigenous Knowledge.* Golden, CO: Fulcrum Publishing.

———. 2013. Introduction: Climate Change and Indigenous Peoples of the USA. *Climatic Change* 120(3): 509–515.

Williams, Brett. 2001. A River Runs Through Us. *American Anthropologist* 103(2): 409–431.

Williams, Terry, and Preston Hardison. 2013. Culture, Law, Risk and Governance: Contexts of Traditional Knowledge in Climate Change Adaptation. *Climatic Change* 120(3): 531–544.

Williams, Walter L. 1979a. Patterns in the History of the Remaining Southeastern Indians, 1840–1975. In *Southeastern Indians Since the Removal Era.* Walter L. Williams, ed., pp. 193–210. Athens: The University of Georgia Press.

———. 1979b. Southeastern Indians Before Removal: Prehistory, Contact, Decline. In *Southeastern Indians Since the Removal Era.* Walter L. Williams, ed., pp. 3–26. Athens: The University of Georgia Press. Wilson, Xerxes. 2013. Terrebonne Parish Presidents Asks for Help in Flood Insurance Fight. *Daily Comet*, August 20. [Available online at www.dailycomet.com/article/20130820/articles/130829935].

Wong, Edward. 2016. Resettling China's 'Ecological Migrants'. *New York Times*, October 25. [Available online at www.nytimes.com/interactive/2016/10/25/world/asia/china-climate-change-resettlement.html?ref=asia&_r=1].

Wood, Mary Christina. 2013. *Nature's Trust: Environmental Law for a New Ecological Age.* New York: Cambridge University Press.

INDEX

Note: Page numbers in *italic* indicate a figure on the corresponding page.